CIVIL ENGINEERING AND ENGINEERING MECHANICS SERIES

N. M. Newmark and W. J. Hall, *editors*

Design
and Planning
of Engineering
Systems

DALE D. MEREDITH *State University of New York at Buffalo*
KAM W. WONG *University of Illinois at Urbana-Champaign*
RONALD W. WOODHEAD *University of New South Wales, Australia*
ROBERT H. WORTMAN *University of Connecticut*

PRENTICE-HALL, INC., Englewood Cliffs, New Jersey

Library of Congress Cataloging in Publication Data

Main entry under title:

Design and planning of engineering systems.

    (Civil engineering and engineering mechanics series)
    Bibliography: p.
    1. Systems engineering. I.–Meredith, Dale Dean.
TA168.D48   620'.72   72-12714
ISBN 0-13-200196-9

© 1973 Prentice-Hall, Inc., Englewood Cliffs, N.J.

Printed in the United States of America

20 19 18 17 16 15 14 13 12 11

Prentice-Hall International, Inc., *London*
Prentice-Hall of Australia, Pty. Ltd., *Sydney*
Prentice-Hall of Canada, Ltd., *Toronto*
Prentice-Hall of India Private Limited, *New Delhi*
Prentice-Hall of Japan, Inc., *Tokyo*

CONTENTS

3

MATHEMATICAL MODELING
OF ENGINEERING SYSTEMS

6

ORGANIZATIONAL NETWORKS   *164*

7

DECISION ANALYSIS   *196*

8

SYSTEM SIMULATION   227

9

SYSTEM PLANNING   258

FOREWORD

The aim of this series of monographs is to make available to students and practicing engineers modern scientific approaches to engineering in the most simple and direct manner possible. This text, prepared for use primarily by undergraduates, was written to introduce the student to the concepts and applications of systems engineering. We expect that with this text the student will gain an appreciation for the continuous interactive process that occurs in planning and design, processes which are often inseparable. Moreover, the student should gain an appreciation that design calls for more than merely involvement with technical and functional aspects but also consideration of societal aspects. It is apparent that in the years ahead engineers must be trained to interact with all modes of planning and design.

We believe that this text, one of the first of its kind, will prove attractive from another point of view, namely that the material is self-sufficient; there is little need for reference to other materials in order to follow the material presented.

We congratulate the authors on the approach they have adopted, and expect that this text will constitute a major contribution to engineering education.

W. J. Hall and N. M. Newmark

PREFACE

The engineer has professional obligations to his client and to society. He must exercise judgment and make decisions for which he will be held responsible. Society has become increasingly aware of and sensitive to the consequential impacts that develop owing to the nature and implementation of engineering works. Therefore, in order to satisfy his obligations to society, the engineer must be able to anticipate a variety of viewpoints and requirements and to incorporate these into his decision-making activities. This calls for more than the mere involvement with the technical and functional aspects of problems. It also requires the consideration of social goals and objectives and an appreciation for the continuous interactive process that occurs in planning, design, and project implementation.

The engineer must perceive problems in their full environmental context and seek the solution that best satisfies the goals and objectives of his client and society. The approach presented in this book leads the engineer to pose problems in greater depth, to focus on the necessary integration of the factors influencing the problem, and to develop objective rationales for his decision making in order to fulfill his professional role in society.

In the design and planning of engineering systems, the engineer must be aware of the environment in which the problem arises; he must establish relevant goals and objectives; he must define and model the problem and alternative solutions; he must provide for the selection and implementation of the best acceptable solution. The degree to which an engineer is able to perform these activities establishes his professional capabilities. It is hoped that this book will help engineering students to develop an awareness of these activities and to develop an approach that will enable them to meet these professional challenges.

The material in the book falls into three classifications. Chapters 1 through 4 are concerned with defining and modeling problems. Because realistic problem definitions and the decision environment are often complex, no single decision process can be adopted by the engineer for all problems. Consequently, Chapters 5 through 11, although containing modeling concepts, emphasize a spectrum of approaches to decision problems. Finally, Chapter 12 hints at how the engineer can integrate into the design and planning process for a complex engineering problem the various concepts and approaches presented.

The book has been designed to be self-contained. Although it has rigor, it presents elementary but powerful concepts that can be applied to complex professional problems. The concepts are emphasized in such a way as to encourage the student to define, model, and solve problems in a creative and professional manner.

The material developed has been successfully taught in a one-semester course at the sophomore level for several years. It could be presented at the junior-senior level or as a two-semester course beginning in the freshman year. Our experience has been that students gain a better understanding of the concepts when they are confronted with developing a professional approach to an engineering problem. This is best done by incorporating the textual material into a semester project.

The problems at the end of each chapter range from simple reiteration of material covered in the text through a series of professional situations that can readily be handled with the concepts presented. To encourage the use of student projects and to suggest synthesis, the problems have been arranged throughout the book to bear on different aspects of a variety of engineering areas. At the end of each chapter these areas are: transportation (problem 3), industrial production (problem 4), high rise building (problem 5), water resources (problem 6), and construction planning and management (problem 7). The problem set associated with any area can be used as a basis for a semester project or as a guide to the presentation of the material in any engineering discipline.

Special thanks and appreciation are due to the American Society of Civil Engineers for granting permission to use the series of photographs included in the book. Many of the projects shown are winners of the ASCE Outstanding Civil Engineering Achievement Award. The annual award is based in part on the engineering project that contributes "to the well-being of people and communities, resourcefulness in planning and in solution of design problems, pioneering in use of materials and methods, innovations in construction, and unusual aspects and aesthetic values."

We gratefully acknowledge extensive debts to our many colleagues who have passed on their comments on the draft versions of the text. Thanks are due to Professors S. J. Fenves, B. Mohraz, R. J. Mosborg, J. P. Murtha, S. L. Paul, A. R. Robinson, G. K. Sinnamon, and W. Tang who have taught various versions of the course and contributed several of the problems. The authors thank Mr. George E. Morris for his assistance in preparing the illustrations.

The many students who managed with our notes in varying stages of imperfection deserve our greatest appreciation.

In a book of this nature there is always a question of what to include and what to leave out. What is included and emphasized here is our view and responsibility.

DALE D. MEREDITH
KAM W. WONG
RONALD W. WOODHEAD
ROBERT H. WORTMAN

1

THE SYSTEMS APPROACH TO ENGINEERING PROBLEMS

Twin towers of the Port of New York Authority's World Trade Center dwarf surrounding buildings in New York City's financial area. The World Trade Center contains offices for government agencies and private firms engaged in international commerce. Physical, social, economic, and political conflicts had to be resolved in planning, designing, and constructing the center. The center was selected as the Outstanding Civil Engineering Achievement of 1971. (Courtesy of the American Society of Civil Engineers)

1.1 INTRODUCTION

Engineers traditionally provide a service function to society by meeting societal needs within available resources and technology. In carrying out this function, the engineer is involved in a continuous and cyclic interaction with society in which new problems arise, alternative designs are conceived, and solutions are implemented. This interaction requires determining what is needed, when it is needed, and how it is attained and involves the engineer in the creative activities of design and planning.

Planning encompasses certain activities that specify how the end product will be achieved. Thus, planning includes preliminary investigation, feasibility studies, detailed analyses and specifications for implementation, manufacture and construction, and monitoring and maintenance. Planning is concerned with formulating the goals and objectives to be attained, consistent with the political, social, environmental, economic, technological, and aesthetic values of the society involved.

Design involves determining the specific form of the end product—its size, shape, properties—and defining the specific emphasis or character of the planning effort that is relevant to the situation; i.e., how much investigation, how many feasibility studies. Engineering design is more often concerned with the detailed specification of the facility components and their interrelationships with one another. It requires consideration of the physical laws of nature and the properties of materials and equipment. In many cases, however, the end product of the design process may be a specialized plan such as a transportation plan, a community development scheme, or the construction plan for a building.

Furthermore, the overall planning activity and professional involvement by the engineer requires consideration of the detailed design aspect of the product. In addition, the design activity in its broadest sense involves determining the extent of the effort and character to be assigned to each step in the planning phase. Thus, the planning and design efforts complement each other.

The design and planning process that the engineer utilizes in solving societal problems is shown in Figure 1.1. This process begins with the recognition of an existing need or issue in the real world that calls for a facility or system of facilities to satisfy the need or issue. The necessary first step in the process is to investigate the environment surrounding the need and extract relevant information and data from the environment in order to help define the problem and establish the problem model. Within the context of the real world, a problem can involve a collection of needs and interests that may conflict with one another. Hence, the engineer may be required to consider the following aspects in relation to

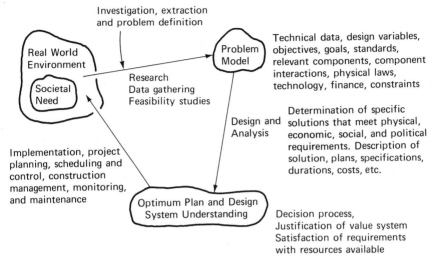

Figure 1.1

The Basic Design and Planning Process

a particular facility: profit, cost, marketability, quality, reliability, performance, life, simplicity, safety, and elegance, as well as political and social acceptability.

Next, the engineer must define a model that represents his concept of the problem. Based on this problem model, the engineer must develop an analysis and design procedure that enables him to define the problem and select the best solution to it. Finally, he must be concerned with the difficulties that are associated with the actual implementation of the conceived solution.

Although Figure 1.1 depicts these efforts in a sequential manner, there may be considerable repetitive cycling within the planning and design process itself. This occurs when:

1. More detailed investigation leads to new insights that require a redefinition of the problem; or

2. Limitations on knowledge and analytical capabilities do not permit the simultaneous consideration of all aspects and implications at any one stage; or

3. Newly discovered technological, financial, social, or political constraints impede the implementation of proposed solutions.

Thus, an iterative process becomes necessary and is illustrated in Figure 1.2.

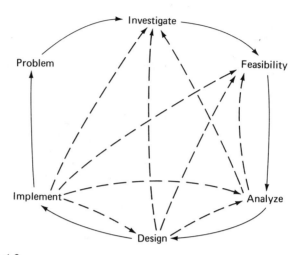

Figure 1.2

The Interactive Nature of the Design and Planning Activities

In meeting his professional responsibilities, the engineer must utilize approaches that recognize the complexity of the problems he faces. These approaches must accommodate any interdisciplinary interaction and provide a rational framework for addressing all aspects that are relevant to the problem. Initially, this would appear to be an overwhelming task, but if comprehensive systematic approaches are utilized, solutions can be achieved that address the broader relevancies of the problems.

1.2 THE SYSTEMS NATURE OF ENGINEERING PROBLEMS

In approaching and solving a problem, one of the initial challenges the engineer faces is to understand the nature of the problem, the environment in which it exists, and the response phenomena that are associated with the problem and the environment. The nature of the problem can provide an indication of the intrinsic factors that are involved in creating the problem; thus, the solution of the problem must also consider these same factors. The environment consists of the setting that contains or surrounds the problem. The response phenomena are indicative of the way that the problem area and its environment will respond to some disturbing stimulus. This stimulus could be a modification that is brought about as the result of the solution to some specific problem that has arisen.

If the engineer is to undertake and achieve a rational approach to the solution of the problem, it is imperative that his definition and understanding adequately represent the problem in its real-world context. Thus, he must recognize the complexity of the real world and accommodate the natural as well as the socio-political forces that are involved. Certainly, there are limitations on the scope of the problem that can be investigated, and these limitations must also be recognized in formulating a solution. This aspect of the problem is an important consideration and is discussed further in the next section.

There are numerous examples that demonstrate the highly complex nature of engineering problems as well as the interaction and response of the problem area and its environment.

1. An interruption in freeway traffic in a city can have an impact not only on the use of that freeway, but also on the city's entire transportation system. By the same token, freeway improvement could affect the entire transportation system.

2. Failure to receive material that is required can delay construction of a facility and result in the temporary unemployment of construction workers.

3. Failure of a machine part can cause a malfunction or limit the operation of that machine such that production is greatly reduced or stopped.

4. Modification of some part of a foundation can have widespread implications on the loads the building can sustain and, therefore, limit the use of the building.

In all cases, the effect of the interaction, response or misfunction may be ultimately traced to the point that social, political, or economic consequences can be defined. Furthermore, all the problems are characterized by the fact that a disturbing stimulus produces an effect on the environment.

This effect is important and interesting to the engineer in two ways. First, the disturbing stimulus defines the cause of a problem; thus the definition of this stimulus permits the engineer to direct his solution at the cause. Second, when a solution to a problem is posed, it becomes the disturbing stimulus, and the engineer is concerned with the consequences that result throughout the affected environment. Part of the professional challenge to the engineer is the fact that his analysis and design must not only consider the immediate problem but also their various consequences. Thus, the engineer must treat the problem and the environment as a whole because of their interactive aspects rather than deal with some fragmented aspect in an isolated context.

A facility with which an engineer is involved is in reality part of some larger entity. In addition, that facility contains a number of parts or components that must be integrated to fulfill some purpose or function. In essence, the engineer is dealing with a system or a system of systems, and the approach that he utilizes must recognize and deal with the problem on this basis.

A system can be defined as a collection of components, connected by some type of interaction or interrelationship, which collectively responds to some stimulus or demand and fulfills some specific purpose or function. In a system, each component responds to stimulation according to its intrinsic nature, but the actual stimulation it receives and its subsequent actual behavior is conditioned by the presence and interaction of the other system components. Therefore, the demands placed on a system call into play the individual behavior of the system components that collectively develop a synthesized composite behavior producing the system's response. In dealing with a system, it is possible to identify the following characteristics:

1. There is some specific purpose or function that must be fulfilled or performed;

2. There are a number of components (at least two) that can be identified as necessary ingredients of the problem. Furthermore, each component has a variety of attributes that implicitly, physically, and behaviorally are necessary for its description;

3. The components are interrelated in some manner satisfying interface consistency between the components; and

4. There are constraints that restrict the system's behavior and the individual component response.

The following examples represent an attempt to identify those characteristics that are listed above for various systems.

Example 1.1   A freeway interchange may be considered as a transportation system.

System purpose: To allow the orderly flow of traffic at the intersection of two highways

System components: Two highways, overpass, ramps, traffic signals, traffic signs, traffic, etc.

System structure: Physical layout of the interchange, traffic flow

System constraints: Traffic volume, human reaction
time, traffic regulations, etc.

Example 1.2   A city building is an engineering system.

System purpose: To provide a place for shops,
offices, and apartments

System components: The physical structure, the floors,
the elevators, heating and light-
ing, etc.

System structure: Use of space, physical layout, etc.

System constraints: Floor-space requirements, safety
features, cost of construction,
surrounding environments, etc.

Example 1.3   A sewage disposal system in a city.

System purpose: To dispose of refuse from city
buildings

System components: The buildings, sewage pipes, processing plants, outlet for processed
sewage, etc.

System structure: Sewage flow, processing, engineering layout, etc.

System constraints: Pollution levels, population density and distribution, etc.

Although the emphasis to this point has been on one system, the engineer may find that the problem actually involves several levels of systems, as illustrated in Figure 1.3. In this case, each level of system must be examined for the components that it contains and the interactions within it as well as with higher and lower levels of systems.

In addition to identifying components, the interactions between the components must be defined. The components and the interactions define the structure of the system. The structure is extremely important because it identifies the manner in which the system behaves when some part of the system is stimulated or modified. In essence, the structure is a key to determining system response.

In the conduct of his work, the engineer may be concerned with

TRANSPORTATION SYSTEM

Vehicle	Residential street	Collector road	Rural road	Interchange	Freeway

Driveway   Intersection   Frontage road   Access road

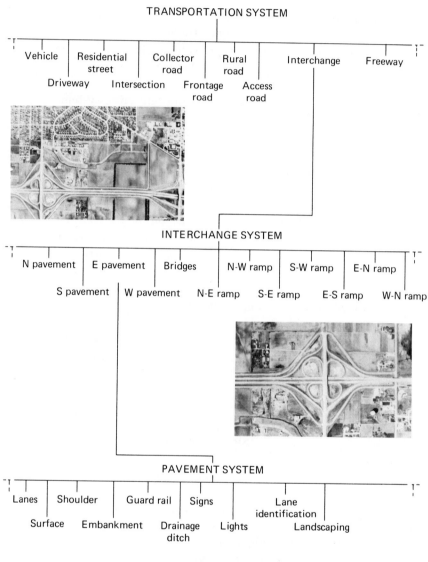

INTERCHANGE SYSTEM

N pavement	E pavement	Bridges	N-W ramp	S-W ramp	E-N ramp

S pavement   W pavement   N-E ramp   S-E ramp   E-S ramp   W-N ramp

PAVEMENT SYSTEM

Lanes	Shoulder	Guard rail	Signs	Lane identification

Surface   Embankment   Drainage ditch   Lights   Landscaping

Figure 1.3

Levels of Systems (Aerial photographs courtesy of Chicago Aerial Survey)

two general types of systems, physical and organizational. His involvement with physical systems is somewhat obvious because engineers are associated with physical entities. In addition to the physical systems, the engineer must determine the sequence of activities that lead to the accomplishment of some project. Furthermore, the engineer must operate within some type of organization in which he might be considered as one component. In this sense, the engineer is equally concerned with organizational systems because they affect the way he operates and the manner in which a project is carried out.

The engineer must capture a representation of the real situation in the problem model that reflects important factors that can be evaluated, manipulated, and prescribed so that he can propose a specific facility or solution that will meet the approval and values of society.

1.3 UNDERSTANDING THE PROBLEM ENVIRONMENT

Prior to developing a definitive statement of a problem, the engineer must first recognize and understand the environment in which the problem is embedded. This is an essential first step in approaching a problem irrespective of its size or complexity, because ultimately the implementation of any solution that he may conceive and propose is contingent on how well he understands the total problem environment. Thus, his failure to adequately define the environment may result in the rejection of his proposed solution. This rejection may be physical in that the facility physically fails; economical, in that it is not feasible; and political and social, in that it is not permissible or acceptable. With respect to a problem, the environment includes the systems that are involved in, and influence the problem and also those systems that are affected by the problem.

As problem complexity increases, the engineer encounters greater difficulty in approaching and defining a problem. He must consider not only specifics relating to his disciplinary area expertise, but must also consider and incorporate the influences and consequences of his solution on the environment in which the problem exists. Consequently, he must utilize an approach that recognizes all the components involved in the problem and the spheres of influence that exist in the environment.

The systems approach is such a technique and represents a broad-based systematic approach to problems that may be interdisciplinary. It is particularly useful when problems are complex and affected by many factors, and it entails the creation of a problem model that corresponds as closely as possible in some sense to reality. Its usefulness increases with problem complexity because it permits the engineer to take a broad over-

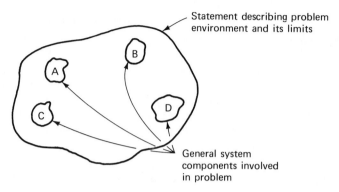

Statement describing problem
environment and its limits

General system
components involved
in problem

Figure 1.4

Problem Environment and Component Identification

all view of the problem under consideration. Thus, a clearer understanding of constraints, alternatives, and consequences that are associated with the problem may be obtained.

To insure that an overall view is attempted, the systems approach discourages the engineer from *initially presenting* a specific problem definition or from *initially adopting* a particular model, mathematics, or solution algorithm. Instead, the systems approach emphasizes that the problem environment be defined in broad terms so that a wide variety of needs can be identified that have some relevance to the problem. These needs should reflect the complex factors, relationships, and conflicts implicit in the problem environment, and which exist in the real-life context, or reality, of the problem.

This initial phase of the systems approach stresses the often neglected fact that the nature of the problem should be critically examined

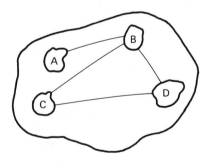

Figure 1.5

Problem Environment and System Components with Structure

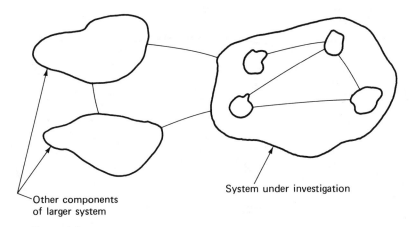

System under investigation

Other components
of larger system

Figure 1.6

System under Investigation as a Component of a Larger System

and explored. In too many situations, a problem statement is too quickly assumed, or too simply specified, so that the validity of a solution is very questionable and may have little or no relevance to the facts of reality. In such cases, a solution that is based on an erroneous problem statement may not solve the problem under investigation or will have unanticipated consequences. These consequences can be of more concern than the original problem because of potential adverse effects on other systems.

Thus, in the systems approach, the initial task is to delineate a general problem environment that permits the general system that is associated with the problem to be identified. Furthermore, the compo-

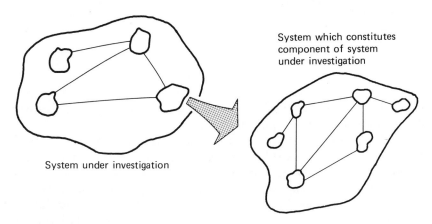

System which constitutes
component of system
under investigation

System under investigation

Figure 1.7

Microanalysis of System Components

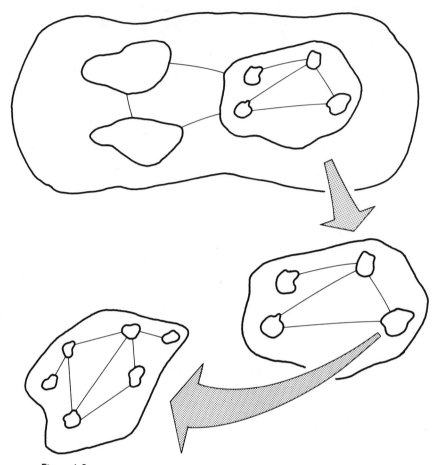

Figure 1.8

Hierarchical System Structure

nents that are contained in the system should be identified. This step is depicted graphically in Figure 1.4.

Interactions with other components must now be determined for each of the components that is identified for the system. At this point, the engineer should be concerned with the fact that an interaction does occur in the functioning of the system, not whether this interaction can ultimately be modeled and analyzed. Taken compositely, the interactions serve to form a system structure that depicts the relation of the system components and their functioning with respect to other components. In essence, the system structure defines the system and influences its behavior. For the hypothetical system shown in Figure 1.4, the system structure has been added and is shown in Figure 1.5.

The engineer must recognize that the problem may involve or be related to other systems. Thus, in defining the problem, these other systems must be identified. For example, a system with which an engineer may be directly concerned is in reality a component of some larger system. Thus, for a comprehensive view of the problem, the engineer also must define the components and structure of the larger system. This relationship of the system under investigation to a larger system is shown in Figure 1.6. Viewing the system under consideration in the context of the larger system has the following benefits. First, a clearer definition of system function can be ascertained; and second, the consequences of modifying the system can be established in the broader context.

A microscopic investigation of the system under study can be made by subdividing each of the components into component parts. This permits a detailed investigation of that component to determine possible design modifications. This microanalysis of the system components is depicted graphically in Figure 1.7. An investigation of each component reveals that it is a system in itself.

Thus, a hierarchical system structure can be determined that permits analysis of the systems at the various levels. This hierarchical structure is shown in Figure 1.8 and compositely depicts the system components that were shown in Figures 1.5, 1.6, and 1.7. This hierarchical structure permits the analysis of the system in terms not only of the system under consideration, but also in terms of both higher- and lower-level systems. Furthermore, this structure provides the framework for analyzing the comprehensive aspects of the problem as well as the technical details that must be addressed and considered.

1.4 DEFINING AND SOLVING PROBLEMS

The preceding section described how to develop an understanding of the general environment in which the problem occurs. This general environment reveals the system or systems that are involved and provides an overall view that may serve to guide the engineer. Certainly, this view of the problem permits the engineer to examine, in a broader scope, the potential alternatives that might be considered in solving the problem and provides a comprehensive judgment with respect to these alternatives. In essence, the problem environment provides the overall framework for attacking and solving the problem.

In addition to determining the hierarchical system structure, each level of the system must be examined and goals and objectives defined. For each level of system, a specific purpose or function can be stated that indicates the needs that must be met by that system. In essence, these goals should reflect the stated purpose and function. Normally, the purpose and function can be ascertained by defining the role of a system in relation to the next higher-level system. The system goal is usually related to the objective of the next higher-level system, as indicated in Figure 1.9.

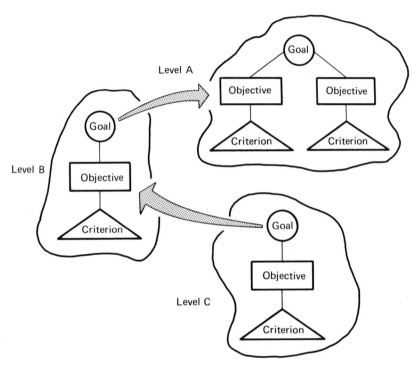

Figure 1.9

System Goal as Defined by Role of System in Relation to Next Higher System

The objectives reflect ways in which the goal can be achieved. For a particular system, the components can indicate the different variables that can be modified in order to fulfill the goal. Ultimately, criteria will have to be defined that indicate how the objectives are to be measured.

In utilizing the systems approach, a basic difficulty lies in determining how much of the problem environment is to be considered in the problem definition. Obviously, the investigation increases considerably with each higher system level that is considered. Also, the solution that is achieved depends on that portion of the environment with which the engineer can actually deal.

Various problem statements or definitions can be generated depending on how many and which system levels and components are incorporated into the engineer's view of the problem. For example, Figure 1.10 depicts two different problem definitions that are the result of in-

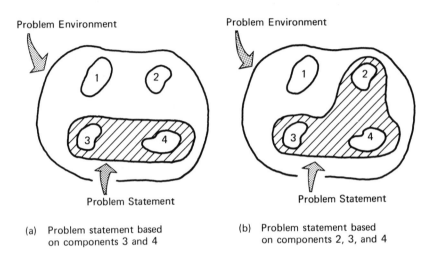

(a) Problem statement based
     on components 3 and 4

(b) Problem statement based
     on components 2, 3, and 4

Figure 1.10

Different Problem Statements Based upon Components That Are Included

cluding different components. In each case, the problem statement represents that portion of the problem with which the engineer can actually deal in seeking a solution. In practice, the breadth and depth of the investigation may be decided in one of several ways.

First, a limit may be reached when some aspect of the problem lies beyond the influence or control of the professional or professionals who are dealing with the problem. This type of limit would normally be encountered as the breadth or scope of the problem increases.

Second, a lack of knowledge concerning a system and its functioning may limit the definition of the problem environment. This lack of knowledge can be related to both the breadth and depth of the investigation. In both cases, the limit is created by encountering a system in which the components, structure, or behavior are not understood. This type of limit can be overcome to a certain degree by including pertinent professionals from other disciplines in the study of the problem. In addition, the inclusion of higher-level systems in the hierarchical structure will require that persons with a higher level of decision-making responsibility be involved in the problem and its solution. These persons may or may not represent the same professional discipline; thus, a problem may result that is related to the transfer of technical knowledge.

Furthermore, the inclusion of lower levels of systems from the hierarchical structure requires greater depth in technical expertise concerning specific aspects of the problem.

Finally, the time requirements for the study may also pose a constraint on both the breadth and depth of the investigation. As some part of the problem is either considered more broadly or in more detail, the time requirements may increase considerably. Normally, the engineer faces some time constraint that dictates when an answer to the problem must be attained. This time constraint certainly must be recognized and considered in defining the scope of the investigation and the resources that are available to accomplish the investigation.

Thus, in addition to determining the hierarchical system structure associated with the problem, the limitations of influence, knowledge or ability, and time also must be defined. From this, a statement of the problem emerges that defines the aspects of the problem environment that will be included in the investigation.

With limits imposed on the study, the engineer must decide how to treat components or other system levels that are outside the constrained or limited problem environment. Several alternatives may be considered, which include the following:

1. When a component provides input or a lower-level system is involved, it may be treated as either a constant or variable input;

2. Where quantifiable relationships cannot be utilized, the influences outside the defined limits may be considered in a subjective manner; or

3. In some cases, components or other levels of systems which lie beyond the defined limits may have to be ignored. This alternative is not desirable, but may be necessary because of the limits that are imposed.

The problem definition, then, will delineate the problem environment and will make clear what goals are to be achieved, what difficulties are to be overcome, what resources are available, what constraints will exist for an acceptable solution, and what criteria will be used to judge the validity of a possible solution.

Having developed a statement of the problem, the engineer must now decide how the problem will be resolved and the manner in which the components are to be considered in this solution. In order to initiate the analysis and design, the engineer must develop a model that represents the problem as defined by the problem statement.

A model is a conceived image of reality; it normally will be a simplification of reality. This does not mean that all models are simple. The complexity of the models depends on the object or process being represented and the purpose of the inquiry. Models may have little resemblance to the actual appearance of the original, but in symbolic terms reproduce the essential elements of reality.

In analyzing a problem, a number of different types of models may be employed according to the requirements of the analysis. These might include:

1. Iconic models, or actual physical representations;

2. Analogue models, or schematic representations of flow processes and dynamic operations; and

3. Mathematical models, which present the situation in mathematical terms.

The analogue type of modeling was utilized in the preceding section in that the schematic representations were used to depict systems and the system environment. In later chapters, there is further discussion of the use of analogue models and of the variety of uses for mathematical models.

Engineers use these models for different purposes, as illustrated by the following general classification:

1. Descriptive Models. The descriptive model is used to present the relationships, order, and sequencing of the systems, and systems components, activities, or analyses with which the engineer is involved on a particular problem. More specifically, the engineer uses descriptive models to describe the manner in which something is accomplished as well as for the detailed specifications of what is involved. Thus, the descriptive model provides the framework for his efforts.

2. Behavioral Models. Behavioral models are used to represent the response of a segment of reality to an initial disturbance. In analysis and design, they are used to design components for a given response or to determine the system response given the properties of the components and the system structure. This response can be associated not only with the physical aspects of the problem, but also with social and political aspects.

3. Decision Models. Decision models are used to select most favorable solutions from among the alternatives that are available according to criteria that the engineer establishes. They

are used to investigate and resolve conflicts and to select the best alternatives and strategies.

In modeling the problem that is defined by the problem statement, the engineer must examine each of the components that are associated with the problem. For each component that is not included in the problem statement, a decision must be made with respect to how or if the component will be included in the analysis. In addition, the relationship and ordering of each component must be made with respect to other components, and the behavior of this interaction must be explicitly defined. If this cannot be accomplished, it indicates that the engineer is unable to solve the problem as it is defined, and the problem statement must be reexamined.

For each system or component, a need model can be developed, as shown in Figure 1.11. The essential feature of a need model is that it

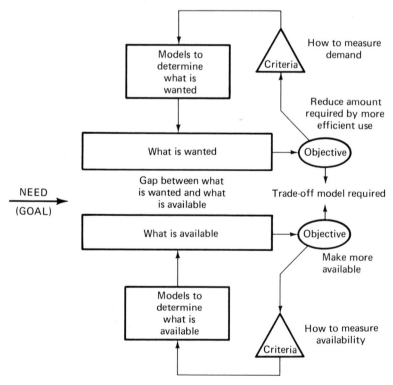

Figure 1.11

Typical Need Model

portrays a disparity between what is wanted and what is available. The resolution of a need implies a reallocation of resources so that what is wanted is made possible. What is available is an indication of the resources that are actually and potentially available for the problem's solution. What is wanted is an unconstrained statement of an ultimate state of the problem's solution. The disparity is an indication that a limited resource problem exists that can be solved with the aid of a decision model. In order to fully develop the need model, the engineer is required to formulate a descriptive model of what is available and what is wanted. In addition, behavioral models will be required to portray the response of the availability and use of resources.

These need models reflect the issues that must be resolved in each component and the manner in which they can be resolved. Thus, for the problem statement, a model emerges that represents how the problem as defined in the problem statement is to be analyzed. Figure 1.12 represents this model in a general form.

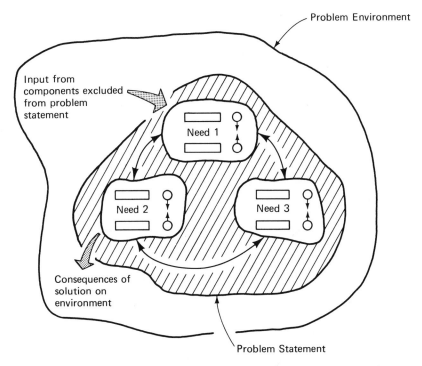

Figure 1.12

General Model of Problem Statement

Defining the problem and modeling the problem statement are basically only part of the overall design and planning process. Having determined a model of the problem definition and its structure, the engineer must acquire necessary data and carry out the analysis and design according to the framework and issues that he has formulated. Although these activities lead to a technical solution of the problem, the task of implementing this solution must be achieved if the design and planning efforts are to be successful.

1.5   AN APPLICATION EXAMPLE: THE HIGH-RISE BUILDING

To demonstrate the application of systems concepts, an example of a systems problem and its definition is presented in this section. The purpose of this example is to demonstrate how systems concepts can be applied, the manner in which a problem can be defined, and the way in which an engineer might proceed in solving a complex problem. The planning and design of a high-rise building has been chosen for this example.

High-rise buildings are very common in large cities because they satisfy the social demand for sheltered space and meet the constraints imposed by the local limited availability of land and its high cost. The high-rise building (HRB) reflects the impact of technology and professional creative concepts in its height, form, and efficient use of materials. (Anonymous, 1972; see bibliography for this chapter.) In addition, it must provide the environment and services necessary to satisfy and attract tenants and customers and must meet financial and investment requirements.

Although the HRB is used as a specific example, it portrays the typical characteristics of an engineering systems problem as indicated by the following:

1. The engineer's involvement is such that he is placed in a position where a number of disciplines are required in order to address the variety of design aspects that must be considered. Thus, the engineering discipline contribution cannot be considered in a totally isolated context.

2. Although engineering technologies play an important role in planning the building, the final design must take into account social, economic and political aspects of the problem. These factors may provide the major rationale for decisions other than those that are made purely on technical grounds.

3. The rational solution of the planning and design problems requires a great deal of discipline interaction because of the interactive nature of the various aspects of the problem itself.

4. An organizational framework is required to resolve disciplinary conflicts and structure the manner in which the conflicts are resolved. The organizational framework must depict the peculiar nature and purpose of the HRB and establish the hierarchical responsibilities of the various disciplines involved.

5. Since the various technologies and disciplines permit a large number of feasible solutions involving form and spatial arrangements, a significant effort is required to formulate the building purpose and function. The purpose and function of the facility serve as the basis for guiding the entire planning and design effort.

6. Finally, the HRB has a large number of components of all types that can be manipulated to achieve a solution that best fulfills the requirements of the facility.

These features and considerations will occur in different degrees in any problem that the engineer confronts, and they must be evaluated and resolved early in the problem formulation.

Traditionally, the engineering and architectural professions have been involved in supplying the housing needs of society. The HRB provides an example where the severely confined space of the facility and the many complex demands of the owner and user-tenant force the professions into intimate contact and interaction during the definition, design, and construction of the building.

Many complex and interacting problems exist in all phases of the building process, from its initiation by a group of investors to its rental occupancy by tenants. For a given ground area, the volume, and hence the available and potential rental floor space, increases with building heights. However, increased building height raises structural, mechanical services, and construction costs, so that eventually increased height meets financial and rental constraints from the owners and tenants.

Providing essential services, such as elevators, stairways, and heating and air conditioning consumes floor space and is directly related to the height, volume, and surface area of the HRB. Maximizing floor space requires improving the insulation and thermal properties of the HRB shell and thus contributes to both the cost and structural and architectural problems. Minimizing elevator service frees floor space but reduces service to tenants and transient customers, thus encountering a consumer constraint. A psychological problem is involved in excessively high buildings, which is aggravated by economies in the number and speed of elevators, as well as by structural vibrations and deflections of the HRB under wind loads. Finally, the overall proportions of the building and its space allocation must be viewed in the light of the owners' expectancy of the life earning capacity of the building in relation to its initial and maintenance costs. In the above illustrative sample of aspects of the HRB, many trade-offs and decisions must be made to resolve the conflicting and interacting demands made on the building. The problems raise issues involving more than technology and can be divided into two broad areas, namely:

1. Internal—those associated with demands on the limited internal space of the building and the associated service and environmental requirements; and

2. External—those associated with the demands created by the HRB on the local political, social, and economic environment in which the building is embedded.

Internally, the problem is to define, design, and provide space that fulfills the needs of those who will utilize the facilities. The allocation of space raises broad questions of whether certain spaces are required and to what extent and economic scale. It requires resolving a number of technological questions and/or a large number of conflicting demands on the space.

Thus, constructing an HRB requires the blending of a number of professional disciplines, with each exercising its unique contributions to the solution. Foundation, structural, mechanical, and electrical engineers are involved, as well as architects and building contractors.

The external problems are generated by the need for space, its recognition, and attempts to provide for its satisfaction. Problems arise in sensing and evaluating the demand and a market analyses may have to be made to estimate this demand. A significant issue involves determining the nature and extent of the constraints such as building codes and zoning laws that will be imposed on the building by local authorities. An assessment must be made of the impact of the HRB on the various city services (water, sewerage, power, transportation, etc.). The constraints imposed represent to some degree an attempt to resolve interfacial problems equitably, so that the surrounding community is not called on to unfairly shoulder service provision costs for the HRB. Procedural problems arise from the legal, political, financial, and construction processes that are called into play during the creative act of producing the building. Finally, the economic feasibility of the entire process regarding investment and the life-use returns to the entrepreneur-owner must be resolved.

As the engineer approaches the planning and design of an HRB, he is not completely free to conceive and resolve alternatives. This fact is due to the nature of the problem and the interactions with other components or elements that are involved in creating an HRB. The engineer achieves a better design when he becomes fully aware of the broader problems and issues. Thus, in any problem such as the HRB, it is imperative that he know the comprehensive elements that control design, the issues associated with each of these elements, and the disciplines or groups responsible if he is to take an active role in the actual design and decision-making process rather than respond passively with purely technical information.

Thus, for the HRB, an examination of the various levels of systems involved and the decision processes required reveals the design role of the engineer. Initially, in devising an HRB, social, economic, and political questions are involved. These questions generally center on the need for space, the marketing potential for such space, the economic feasibility of providing such space, and the political controls that govern the development.

At this problem level, the owner-entrepreneur must establish the range of financial involvement required to provide a variety of HRB solutions related to the size and character of prospective tenant demands for each of the HRB solutions that he can provide. He must deal with bankers and strive to establish financial flexibility so that he can optimize the financial return by selecting the best HRB solution. He must obtain professional advice on the number and types of tenants best suited for his venture and establish attractive rentals that insure his desired return. In addition, governmental control in the form of building codes or zoning laws may be a factor in each of these steps. Figure 1.13 graphically

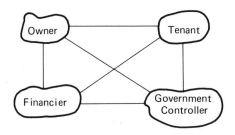

Figure 1.13

Agents Involved in the Decision-making Process

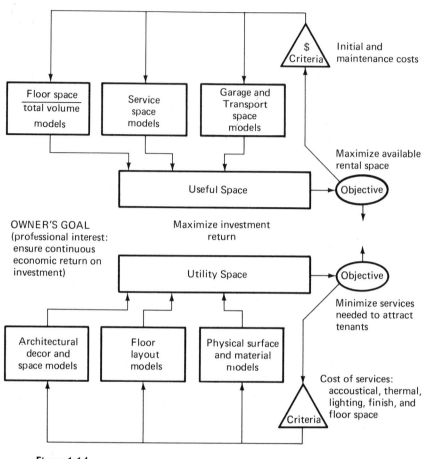

Figure 1.14

Owner's Need Model

depicts these relationships. Although persons or groups are shown in the figure, they represent components or elements at this system level.

For each of these elements a need model can be developed and used as the basis both for developing the explicit nature of the system inter-action or for exposing the basic issues, and for developing the models and analytical methods that can be used in resolving conflicts that arise in the need models. A typical need model for the owner is shown in Figure 1.14 and for the tenant in Figure 1.15.

The owner's goal, for example, may be to maximize the return on his investment. Thus, as far as the internal problems are concerned,

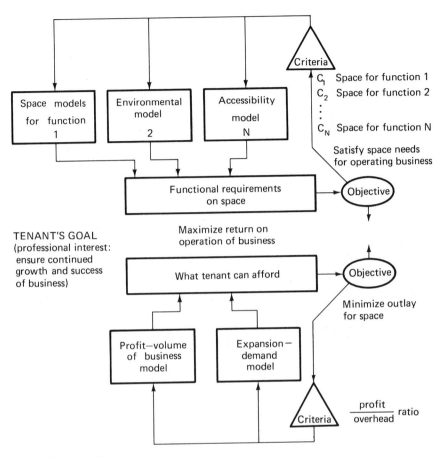

Figure 1.15

Tenant's Need Model

he has, on the one hand, the objective of maximizing the available useful and rental space and, on the other hand, the external problem objective of minimizing the services needed to attract tenants. His criteria for measuring the success of meeting objectives may be the relationships between costs (life, initial, and maintenance) and life rental potential and the cost of services (acoustical, thermal, lighting, surface finish, floor space, unit costs, etc.). He is interested in floor space in proportion to total volume models; service space models, including garage space and transportation; architectural decor models; and floor layout models. In this situation, he needs to resolve conflicts between the various objectives

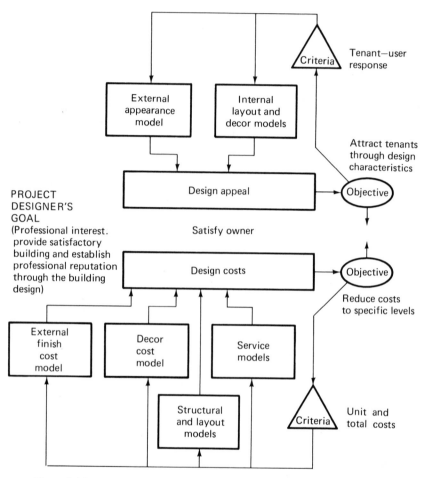

Figure 1.16

Project Designer's Need Model

to best meet his goal. To succeed, he must establish a value scale (criterion) with which to measure objective achievement, develop models representing the various properties of the proposed building, and solve the models to determine the optimal choices.

The tenant's goal may be to maximize the return on the operation of a business, in which case he may approach the HRB from the point of view of minimizing the outlay for space consistent with satisfying the space needed for operating his business. From his point of view, he also must establish goals, objectives, criteria, and technological models.

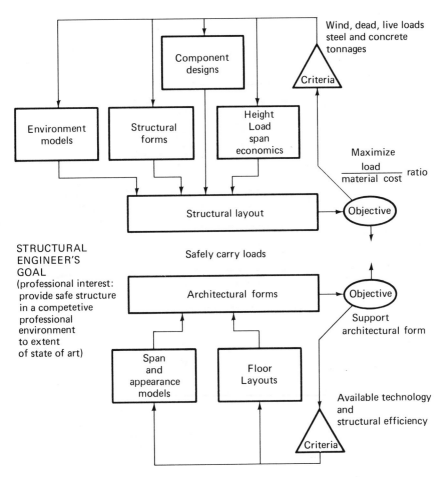

Figure 1.17

Structural Engineer's Need Model

At this level of the problem, the emphasis is on establishing broad feasibility concepts for providing space and its required characteristics. Having made these initial decisions to proceed, the owner must now involve the professional disciplines that are required for the planning and design process. The nature of the building and its components will determine the professional disciplines that are involved. The role of these disciplines with respect to other disciplines is contingent on how much emphasis is given that component in the HRB for which their discipline is responsible. The manner in which interprofessional decisions

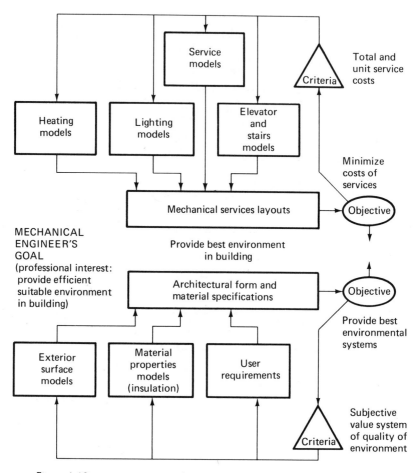

Figure 1.18

Mechanical Engineer's Need Model

are resolved relates directly to the purpose of the HRB and the relative magnitude and importance of the component involved. Because of the number of disciplines involved, the owner does not attempt to deal with each individually, but rather selects a coordinating agent through whom he deals with the various disciplines.

    For example, the following situations demonstrate the different organizational structures that might result.

    1. For the typical building problem in which many alternatives

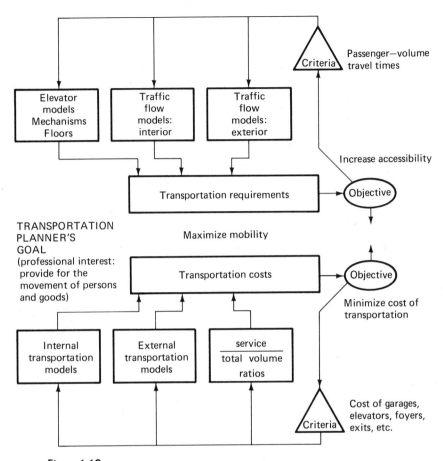

Figure 1.19

Transportation Planner's Need Model

exist, the architect normally serves as the coordinating discipline and interacts with the owner.

2. The structural engineer may be the coordinating agent when the structural support aspects are the dominant problem.

3. For a specific building function that is unique, such as a complex parking garage, the transportation engineer will be dominant because of the critical traffic consideration.

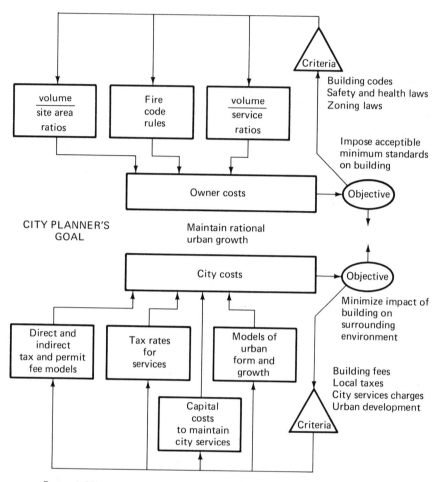

Figure 1.20

City Planner's Need Model

4. For expedient design and construction of a building, the situation may dictate the use of a design-build concept in which the building contractor assumes the leadership role.

The leadership role and the resulting relationships with other disciplines provides the system structure for component interaction and the way in which the disciplines operate and resolve conflicts.

For each component and the representative discipline, a need model

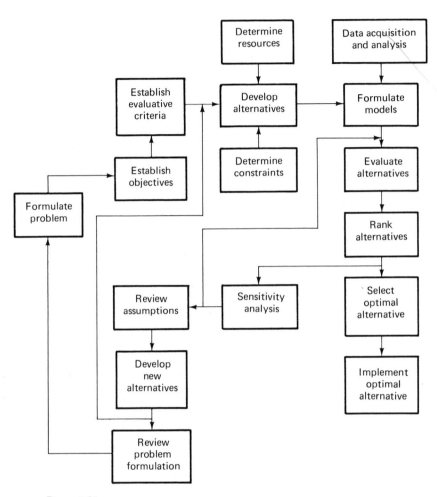

Figure 1.21

Steps in Problem Solution

can be developed. Figures 1.16, 1.17, 1.18, 1.19, and 1.20 indicate need models for various disciplines. Note that the project designer, structural engineer, mechanical services, and transportation goals are related to objectives of the owner, who represents the next higher-level system. The goals of the city planner are directly related to the objectives of the government controller.

For planning and designing the HRB, these concepts can be further developed to other system levels and disciplines. Lower levels are characterized by increasing professional specialization and focus on behavioral, information, and data systems.

The engineer must recognize the system level at which he is working and determine the scope of influence and knowledge limitations that are imposed. If this is done, he can propose design alternatives, develop relevant technical and trade-off models, and justify the basis for his decision-making in relation to the problem. The steps that the engineer might follow in seeking solutions for the problem are outlined in Figure 1.21.

Throughout the remainder of the book, system models that the engineer will find useful in solving system problems at various levels are proposed and discussed.

1.6 PROBLEMS

P1.1. For each of the following situations identify the systems of physical objects, agencies, and social groups involved and the possible interaction conflicts that should be considered in the engineering design and planning process (see Section 1.2).

1. A city council has passed an environmental act requiring that all existing and future utilities be placed underground.

2. An area within a high-density city has an extremely high crime rate, and the city engineer has been asked to determine if there are engineering solutions that can help to reduce the incidence of crime.

3. An underpass within a business district is frequently flooded during storms.

4. A company is changing its product line and desires to remodel its factory to accommodate this change while old products are being phased out.

5. An urban renewal plan requires the demolition of a high-density residential area and its conversion into a recreational use facility.

6. A developmental group desires to develop a recreational resort in a publically owned primitive area and must meet aesthetic, ecological, and environmental constraints.

P1.2. Develop need models for the various levels and decision-makers involved in the situations described in P1.1 (see Section 1.4).

P1.3. Observe a major street intersection in your city. For this intersection, identify the hierarchical system that is involved and the components that are contained in each level of the system. Determine the purpose or function at each level for each system and its components. If you detect problems at this intersection, indicate how an examination of the hierarchical system could be utilized to identify the cause of the problems and possible solutions that might be considered. How is the solution approach for this situation different from one that would be useful for an airport, rail station, large parking lot, or shopping center.

P1.4. Visit a production facility and identify one or more systems that exist. List the components of each system and their purpose or function. To what extent does the system change when the engineer faces the following problems:

1. Spatial layout of material flow;

2. Sequential operations in the product process line;

3. Choice of inventory level for raw materials and finished product in order to meet order and delivery requirements;

4. Effect on product line flows of introducing a new product requiring the use of both general purpose and special purpose machines.

P1.5. Develop structured general problem statement models for a high-rise building in terms of need model components and the decision hierarchy, and hence, develop the relevant goals, objectives, and criteria for the following problems:

1. Selecting floor layout;

2. Selecting floor heights as influenced by floor use, desired artificial environments supplied by utilities, costs, attractiveness to tenants and customers;

3. Selecting material for the building's surface;

4. Selecting the transportation system in the building;

5. Determining size and location of the high-rise building;

6. Establishing zoning requirements for city areas.

P1.6. Figure P1.6 shows the general land use for a river system. Suggest a possible system that can be used to alleviate the problems associated with water supply, flooding, and recreation for the region. Identify the various social, economic, political, and technical issues and conflicts involved in your proposed system.

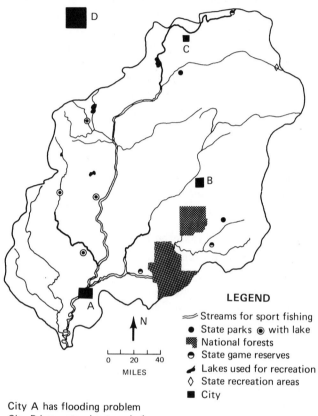

LEGEND

〰 Streams for sport fishing
● State parks ◉ with lake
▓ National forests
◓ State game reserves
↙ Lakes used for recreation
◊ State recreation areas
■ City

City A has flooding problem
City B has water shortage during summer
State parks and recreation areas are
  overcrowded in summer months
City D is large city near basin
Rest of basin is mainly in agricultural use

Figure P1.6

P1.7. Identify systems, system components, structure, and possible constraints in the following situations:

1. A concrete highway bridge spanning a small country stream, and

2. A steel highway bridge spanning a coastal estuary with a swing span to permit shipping traffic.

Can you identify various levels of systems associated with the physical objects, the functions being performed, and the processes that permit, produce, use, and maintain the facilities involved.

P1.8. The hierarchical system structure depicted in Figure 1.8 permits the analysis of systems at various levels. This structure provides the framework for analyzing the comprehensive aspects of the problem as well as the technical details that must be addressed and considered. The problems raised at different levels require different types of answers, models, and decision processes.

1. What kinds of answers and models can you identify at different levels?

2. What degree of detail would have to be incorporated into these models?

3. What kinds and amounts of data might be required at different levels?

2

LINEAR GRAPH MODELING AND ANALYSIS OF SYSTEMS

The Intercontinental Ballistic Missile (ICBM) was selected as the Outstanding Civil Engineering Achievement of 1962. In getting the ICBMs into the air, significant developments were made in the fields of soil mechanics, engineering mechanics, structural design, and underground construction. The methods developed for predicting soil motions resulting from detonations, for designing structures to withstand high over-pressure, and for designing shock insulation systems can be applied in the design of buildings and dams. (Courtesy of the American Society of Civil Engineers)

2.1 NETWORK SYSTEMS IN ENGINEERING

A common feature of many engineering systems is that they are composed of a number of components physically interwoven in the form of a network. Typical examples are the interstate highway system; a city sewage collection and disposal system; a local telephone system; the national transportation system; a city water supply system of dams, pipelines, and reservoirs; and the steel frame of a skyscraper.

In addition, there are many engineering decision problems and organizational systems that, although they do not have the physical appearance of networks, can be usefully interpreted as such; e.g., the flow of decisions and authority within an industrial firm can be described by a network system. Similarly, the schedule of jobs in a construction project may be viewed as a network of activities. The flow of cash among the agencies within a state government or among the departments within a construction company may be considered as a network of cash flow.

A common feature of these decision and organizational examples is that they are viewed in such a way that inherently logical or procedural aspects are used as a means of identifying components and structuring them together into a network system. These networks can be depicted and analyzed through the use of linear graphs; and, in this respect, linear graphs are a valuable aid to the engineer in seeking solutions to problems that involve network systems.

2.2 GRAPHICAL MODELING OF ENGINEERING SYSTEMS

A network system may be graphically represented by a set of points, together with a set of lines connecting some of the points. A simple form is one in which system parameters or components are represented by points (called *nodes*). The interrelationships or connections that exist in the system between the parameters or components are represented by lines (called *branches*) joining the relevant graphical nodes. The graph itself is called a *linear graph* model of the system.

Linear graphs are not drawn to scale, and the relative positions of the nodes and branches do not necessarily represent the actual relative positions of the system parameters or components. Therefore, linear graphs should not be confused with geometric graphs and the physical properties of lines and points. Linear graph theory is an abstract concept and is called *linear* because of its concern with the connectivity of lines and nodes. The primary purpose of a linear graph is to present a concise, graphical representation of the interrelationships of the system that are relevant to the system problem at hand. It thus provides a conceptual way of portraying the environment or system in which a problem is embedded.

The modeling capability of linear graphs is shown in the following examples:

Example 2.1 Consider the problem of a construction contractor who is choosing a route to move some heavy construction equipment from Chicago to Urbana, Illinois. He has no preference between two-lane or four-lane highways, but he does want to avoid traveling on secondary roads because of load limitations. He wishes to find the route that requires the shortest travel time. After studying the highway map and consulting the highway authorities, he has narrowed his choices to three, as indicated in Figure 2.1a. Figure 2.1b is a linear graph model of the three feasible routes.

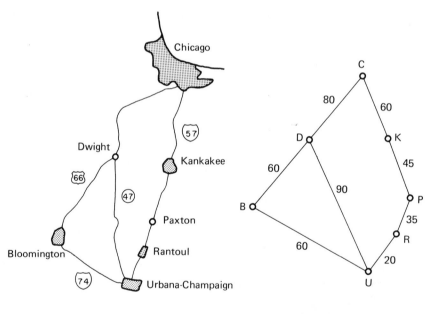

(a) Highway Network Map   (b) Linear Graph Model

Figure 2.1

Linear Graph Representation of a Highway Network

The nodes denote the cities and towns, and the branches, the highway connections. The numbers along the branches indicate the driving time in minutes. From this graph, he easily arrives at the conclusion that his best route is Chicago-Kankakee-Paxton-Rantoul-Urbana, which takes 160 minutes. The second best choice is the Chicago-Dwight-Urbana route, which requires 170 minutes. The route going through Bloomington requires 200 minutes.

A linear graph is constructed with the specific purpose of helping to solve a system problem. Therefore, the choice of parameters to be represented in the graph must depend on the nature of the problem and the system objectives. Suppose that an engineer is preparing a plan for improving the highway network in the same area as is illustrated in Figure 2.1. His graph model may well include the following additional features: feasible routes for new highways, existing highways for which additional lanes can be added, towns that require better highway connections because of industrial developments, recreational sites, etc. He may also choose to use several different symbols among the nodes to differentiate towns, cities, and recreational sites, as well as different symbols among the branches to differentiate the different classes of highways.

Example 2.2   Figure 2.2 is a linear graph model of the pipe network in a sewerage system. The arrows indicate the directions of flow,

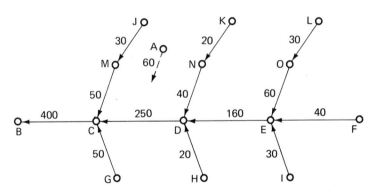

Figure 2.2

Linear Graph Model of a Sewerage System

the nodes represent the junctions of pipes, and the number on each branch denotes the flow capacity in cubic feet per second (cfs) for that section of pipe. Suppose that a subdivision is being developed at A and a sewage flow capacity of 60 cfs is required. A basic problem is therefore raised by the question "How can subdivision A be connected to the existing sewerage network?"

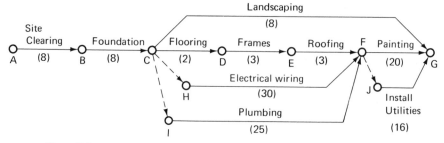

Figure 2.3

Linear Graph Model of a Construction Plan

Example 2.3 Figure 2.3 is a linear graph representing the scheduling problem in assembling a prefabricated house. Each arrow models a job and is labeled with its estimated duration in hours. The directions of the arrows indicate the order in which the jobs must be performed. For example, frames cannot be erected until flooring is completed, but can be started before completing electrical wiring and plumbing. Thus, the model clearly shows the order or precedence for all the tasks and the length of time needed to complete each task. Using such a diagram, the total project duration and the amount of extra or slack time available for each task can be computed. In this example, the shortest possible time the project can be completed is 66 hours, or $8\frac{1}{4}$ days. Such a diagram is called the CPM (Critical Path Method) diagram and is being used extensively for planning and controlling construction progress.

Example 2.4 Consider the problem encountered by a contractor who is performing some work in a river flats area that has been subjected to high water conditions and occasional destructive flooding in the past. There is a four-month period during which he has no use for the equipment either on this job or on others. He can keep the equipment on this job in the river flats, or else move it out, store it, and then move it back, at a total cost of $1,800.

If he keeps it in the river flats, he has the option of building a platform for the equipment at a cost of $500, which will protect it against high water, but not against a destructive flood. The damage that would be caused by high water amounts to $10,000, if there is no platform. A destructive flood would entail a loss of $60,000, regardless of whether or not he builds a platform. The probability of high water in the four-month period is 0.25; the probability of a destructive flood is 0.02. The contractor has to choose from three possible options:

1. Move equipment,
2. Leave equipment in the area and build a protective platform, or
3. Leave equipment in the area unprotected.

Figure 2.4 is a linear graph model of this decision problem.

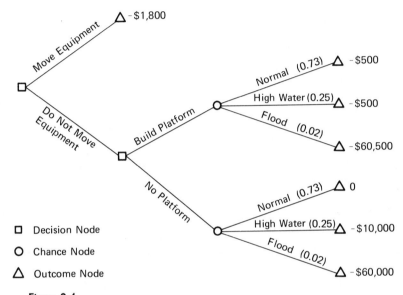

Figure 2.4

Linear Graph Model of a Contractor's Decision Problem

This type of linear graph model is called a decision tree. A branch represents either an alternative available to the decision-maker or a possible outcome that can result from an uncontrollable chance event. A node denotes either the occurrence of an event at which a decision must be made, the occurrence of an event where outcome is determined by chance, or a final condition.

The decision nodes, denoted by squares, are those at which the decision-maker must choose one course of action from among the alternatives then available to him. The chance nodes, indicated by circles, are those at which the decision-maker has absolutely no control. The actual specific outcome from a chance node, of course, depends on nature but, in the prior situation for the contractor, is modeled in Figure 2.4 as a probability outcome using past observations of nature. The numbers on the chance outcome branches indicate the probability that these outcomes would occur. The final outcomes, shown as triangular nodes at the tips of the tree, show the cost to the decision-maker for each possible combination of events.

The preceeding examples of the modeling of physical, organizational, and sequential decision systems problems illustrate the use of linear graphs as an aid in modeling complex problems. The actual modeling of problems using linear graphs requires creativity and an intimate knowledge of the problem on the part of the investigator. The process of applying linear graphs to systems problems might be described as follows:

1. As with other systems problems, the first step is to identify the problem and the system components, structures, and attributes that are involved.

2. The problem must now be represented as a model according to some representational form and convention. In linear graph modeling, the model is limited to the use of nodes and branches as the representational forms. In addition, attributes must be assigned to the branches and nodes.

3. The problem must be posed in terms of a graphical structural property that exists in the model.

4. A linear graph analysis procedure must be established that will enable the graphical property to be processed through the linear graph model.

5. The graphical solution must be transferred back into the systems structure of the actual problem.

In the remaining sections of this chapter, the various aspects of the application and use of linear graph models and analysis are presented and discussed.

2.3 LINEAR GRAPH ANALYSIS: A PATH PROBLEM

Graphical properties relate to the many different structural relationships that can be imagined as existing between the various nodes and branches of a linear graph. The development of these properties permits graph theory to be applied to many engineering problems. As an illustration, several examples that depict graphical properties as related to specific problems are presented in this and the following section.

Consider, again, the contractor's problem involving the movement of heavy equipment over a road network between two towns (see Example 2.1, Figure 2.1). The contractor is interested in routes and is confined to travel along a sequence of roads passing through towns. In addition, although he knows that the roads carry two-way traffic, he only wants to go from Chicago to Urbana, so that a directed one-way route concept results and Figure 2.1 should be changed to portray a directed graph. A directed graph is one in which at least one branch is assigned a unique direction. The branch direction is usually shown by an arrowhead. This discussion raises the need for two graphical properties describing routes; namely, *chains* and *paths*. A chain is a sequence of alternating nodes and branches between two terminal nodes where successive branches have one node in common. A path is a chain in a directed graph where, in addition, all the branches in the path point in the same direction. The concept of a chain is also important from the point of view of system models. Figures 2.1 to 2.4 show linear graph models of various systems; the graphs are continuous, they are not fragmented. In graph theory, this means that the graphs are connected graphs. A connected graph simply requires that every pair of distinct nodes can be joined by at least one chain in the graph. If the graph contains an isolated portion, then it is a disconnected graph.

The linear graph analysis problem then reduces to enumerating paths directed from node C to node U in the directed graph of Figure 2.1b. Because no one-way highway segment exists, the individual chains and paths are identical. The solution process simply requires that the three paths be ranked according to the contractor's criterion of minimum travel time.

Consider the CPM linear graph model of Figure 2.3. It is a directed graph since technological and organizational conditions establish the various conditions before an activity can start. Because all the activities must be completed in specific directed sequences, it is necessary to *traverse all* the seven paths joining the start node A and the end node G. Consequently, if the problem concerns determining the minimum project duration, the graphical structural property forming the basis of the analysis is again the path.

The total time to complete each path is as follows:

Path	Time Required for Completion
A-B-C-G	24 hours
A-B-C-D-E-F-G	44 hours
A-B-C-D-E-F-J-G	40 hours
A-B-C-H-F-G	66 hours
A-B-C-H-F-J-G	62 hours
A-B-C-I-F-G	61 hours
A-B-C-I-F-J-G	57 hours

Thus, the longest path takes 66 hours. This is the minimum time required to complete the project. If there is a delay in completing any of the jobs along this critical path, the entire project will be delayed. On the other hand, if the duration of the project is to be shortened, one or more of the jobs along this critical path must be shortened. Therefore, this critical path controls the duration of the project.

A similar method of analysis can be used for the problem in Figure 2.4. The contractor has the choice of three possible decision paths:

1. Move equipment;
2. Do not move equipment, but build a protective platform;
3. Do not move equipment and build no protective platform.

The consequence of the first path is a definite loss of $1,800. However, the consequences of the other two paths depend on chance. Without a definite knowledge of future events, the contractor must estimate what he can expect to gain or lose in taking path 2 or 3. These estimates provide a single measure of value for the paths with uncertain outcomes. Taking into consideration his available capital and his willingness to assume risk, he must decide on which path he will take. The methods of decision analysis are discussed in Chapter 7.

2.4 LINEAR GRAPH ANALYSIS: A NETWORK FLOW PROBLEM

A large class of engineering networks can be viewed as capacitated flow problems in which one or more commodities (e.g., traffic, water, information, cash, etc.) can be considered as flowing through the network whose branches have various constraints and flow capacities.

Consider, for example, a two-lane, two-way highway network connecting a metropolitan area with a residential suburban town, as shown in Figure 2.5a. Along each road branch is a vector containing the relevant attributes of the road it represents: road identification number, mileage, and flow capacity in one direction in vehicles per hour (vph). Thus, the road between nodes 1 and 3 is identified as road 1, which is 15 miles long and has a maximum capacity of 700 vph in the direction from node 1 to node 3. The two-way capacity would be twice this value.

The engineer is interested in determining the maximum flow, or capacity, of the network during the morning and evening peak traffic hours. Assuming that the morning and evening flows follow identical characteristics, but in opposite directions, the study can be confined to either the morning or evening conditions; e.g., only the flow from node 5

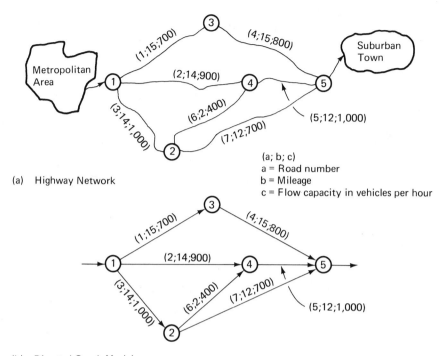

(a)   Highway Network

(a; b; c)
a = Road number
b = Mileage
c = Flow capacity in vehicles per hour

(b)   Directed Graph Model

Figure 2.5

Linear Graph Model of a Highway Network

to node 1 or the flow from node 1 to node 5 needs to be studied, not both.

Consider the evening peak traffic. Each path joining node 1 to node 5 represents one possible route of travel. The first step is to assign directions to the flow in as many branches as possible. The direction of flow in roads 1, 2, 3, 4, 5, and 7 must be from node 1 to node 5 and can, therefore, be assigned. However, the flow in road 6 can be from node 2 to node 4 or from node 4 to node 2, and there may be no *a priori* way of knowing the correct direction to assign to the flow in this branch. Therefore, the problem must be analyzed for both possible conditions. For the first case, assume that the flow is from node 2 to node 4 and thus, the direction of flow in each branch is shown in the directed graph of Figure 2.5b.

If the engineer is interested in determining the flow of vehicles between two intersections that are connected by a highway segment, he can place a pneumatic tube across the roadway and connect it to a counting device. The road tube may be placed anywhere along the highway segment, the only requirement being that the path between the two inter-sections is traversed. Figure 2.6a depicts this situation in terms of a linear

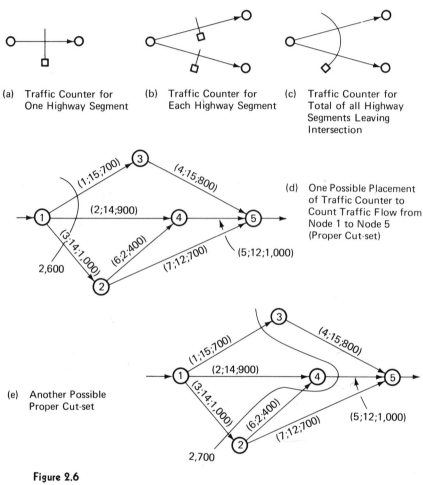

(a) Traffic Counter for
     One Highway Segment

(b) Traffic Counter for
     Each Highway Segment

(c) Traffic Counter for
     Total of all Highway
     Segments Leaving
     Intersection

(d) One Possible Placement
     of Traffic Counter to
     Count Traffic Flow from
     Node 1 to Node 5
     (Proper Cut-set)

(e) Another Possible
     Proper Cut-set

Figure 2.6

The Development of the Cut-set Concept

graph. The intersections are represented as nodes, the roadway as a branch, and the counting tube as cutting the branch between the nodes. If roadway traffic is flowing at its maximum rate, the value obtained from the count represents the capacity of flow between the two nodes. Furthermore, if two roads are involved, two counters could be utilized that would determine the flow on each. Thus, the total flow on both roadways could be ascertained by adding the values obtained from each counter. This situation is shown in Figure 2.6b, in which each of the two branches are cut by the counting device. Again, the important consideration is that counting devices cut across the paths of travel or the branches.

The engineer is interested in obtaining the value of total maximum flow; therefore, instead of two counting devices and two tubes, he really wants to determine the flow that could be measured by one counter and a tube that crosses both roadways. Figure 2.6c represents this concept and shows that the tube across both branches would yield the total flow on those branches.

In the case of the network shown in Figure 2.5, determining maximum flow from node 1 to node 5 would require that a tube be placed across the roadways such that node 1 is on one side of the tube and node 5 is on the other. Figure 2.6d shows one possible placement of the tube. Since roads 1, 2, and 3 all carry traffic flow from node 1 (source) side of the tube to node 5 (sink) side, the total flow capacity from the source side to the sink side is thus the sum of the flow capacity of these three roads; i.e., $700 + 900 + 1,000 = 2,600$ vph. In the terminology of linear graph, branches 1, 2, and 3 constitute a proper cut-set. A *proper cut-set* is defined as a minimal set of branches in a connected graph that, if removed, would break the graph into only two subgraphs. Thus, the proper cut-set in Figure 2.6d breaks the graph into two subgraphs: one containing only node 1; and one containing nodes 2, 3, 4, and 5 and branches 4, 5, 6, and 7. Branches 1, 2, 3, and 6 constitute a cut-set, i.e., their removal separates the graph into two subgraphs. However, they do not constitute a minimal set because only branches 1, 2, and 3 need be removed to form two subgraphs.

Maximum Flow-Minimum Cut Theorem

The above proper cut-set simply indicates that the maximum network flow capacity cannot be greater than 2,600 vph. At this point, the engineer is not certain that he has found the proper cut-set that indicates the maximum network capacity flow. For example, if he cut branches 4, 5, and 7, the total potential flow would only be 2,500 vph. This proper cut-set further indicates that the network capacity cannot be greater than 2,500 vph.

In computing the values of the proper cut-sets, consideration must be given to the direction of the branches included in the cut-set. For example, the cut-set in Figure 2.6e has branches 1, 3, and 5 going from the source side to the sink side, and branch 6 going from the sink side to the source side. The maximum capacity flow from the source side of the cut-set to the sink side should be $700 + 1,000 + 0 + 1,000 = 2,700$ vph. That is, the branch that crosses the cut-set from the sink side to the source side has zero value. In the physical sense, if the pneumatic tube were placed in that position to measure the flow capacity from the source to sink, the tube should not count the cars traveling from node 2 to node 4.

For total network analysis, all the possible proper cut-sets that separate the source from the sink must be considered. Figure 2.7 depicts

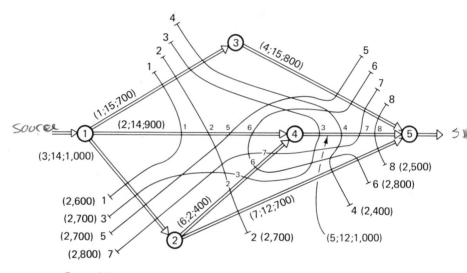

Figure 2.7

Road Network Proper Cut-sets

all the possible proper cut-sets for this particular road network. As can be seen, there are eight possible proper cut-sets that separate the source node 1 from the sink node 5. The components and capacities of the cut-sets are also tabulated in Table 2.1. An entry of $+1$ denotes a positive flow (from source side to sink side); an entry of -1 denotes a negative flow (from sink side to source side); and a blank entry (or zero) indicates that a branch is not included in the cut-set. The maximum network flow capacity is thus determined by the proper cut-set that has the smallest total capacity; in this case, it is cut-set No. 4, which has a capacity of 2,400 vph. Therefore, the maximum network flow capacity between nodes 1 and 5 is 2,400 vph.

Table 2.1

ROAD NETWORK PROPER CUT-SET CAPACITIES

HIGHWAY		CUT-SET NO.							
No.	Capacity	1	2	3	4	5	6	7	8
1	700	1	1	1	1				
2	900	1	1			1	1		
3	1,000	1		1		1		1	
4	800					1	1	1	1
5	1,000			1	1			1	1
6	400		1	−1			1	−1	
7	700		1		1		1		1
Cut-set capacity		2,600	2,700	2,700	2,400	2,700	2,800	2,800	2,500

This analysis approach illustrates the maximum flow-minimum cut theorem, which states: For any source-sink connected network, the maximum flow capacity from source to sink is equal to the minimum of the proper cut-set capacities for all the proper cut-sets separating the source from the sink.

In the example above, a solution was obtained by finding and evaluating all the different proper cut-sets of the graph. In large networks, this approach is both laborious and uncertain because of the difficulty of insuring that all proper cut-sets have been found. The maximum flow-minimum cut theorem can be used to provide upper and lower limits to the network flow capacity, as described in the following paragraphs.

Let C be the network flow capacity. Then the initial cut-set 1, which has a value of 2,600 vph, immediately establishes the relationship

$$C \leq 2{,}600 \tag{2.1}$$

Again, with cut-set No. 8 evaluated at 2,500 vph, there results

$$\begin{matrix} C \leq 2{,}600 \\ C \leq 2{,}500 \end{matrix} \quad \text{i.e., } C \leq 2{,}500 \tag{2.2}$$

As each cut-set is found and evaluated, it plays a part in determining whether the current tentative value of the upper limit for the network capacity must be lowered.

Suppose now that instead of seeking to evaluate all the remaining proper cut-sets, an attempt is made to establish actual path flows through the network. Introducing 700 vph along the path produced by highways 1 and 4 establishes a lower bound for the network capacity as:

$$700 \leq C \tag{2.3}$$

The addition of 900 vph along the path produced by roads 2 and 5 and of 700 vph along the path produced by roads 3 and 7 establishes for all three path flows

$$2,300 \leq C \tag{2.4}$$

Hence, the network capacity can be bounded from both sides

$$2,300 \leq C \leq 2,500 \tag{2.5}$$

Therefore, if an actual flow can be found equal to the minimum value of the several cut-sets found at this stage, it is not necessary to locate and evaluate the remaining proper cuts since they cannot have capacities less than the flow already established. In determining the lower limit on the flow capacity a branch can be in only one path.

In order to determine if 2,400 vph is the maximum flow in the network, the flow from node 4 to node 2 in road 6 must be analyzed. Figure 2.8a shows the directed graph for this condition. The cut-set analysis for this case is shown in Figure 2.8b. In this case the maximum flow capacity is only 2,300 vph. Thus, *the maximum flow will occur in the network with the flow from node 2 to node 4 on road 6.*

For flow networks that have many branches for which the flow directions cannot be assigned *a priori*, the graphical cut-set analysis becomes very long and tedious. It may be more convenient to use one of the procedures described in Sections 2.5 and 3.5.

Once the road network flow capacity has been established and the constraining cut-set identified, an analysis can be made for flow capacity improvement. Since the network flow capacity is limited by the capacity of the roads forming cut-set No. 4, road improvement elsewhere is not effective at this stage. Therefore, at least one of the cut-set No. 4 roads must be increased in capacity for any overall network improvement.

Road 1 can be improved alone, but only optimally for the additional capacity of 100 vph, because the path flow through it then matches the road 4 capacity of 800 vph.

Either road 5 or 7 can be improved by an amount of 200 vph, at which state the new cut-set No. 4 capacity equals the cut-set No. 1 capacity of 2,600 vph. Which of the two roads should be improved will depend on other considerations, such as length of road and costs. A typical problem will be considered later (see Sections 2.6 and 3.5).

The use of the maximum flow-minimum cut theorem for this problem is interesting because it illustrates a problem solution based on the interaction of two different linear graph analysis methods. Although both

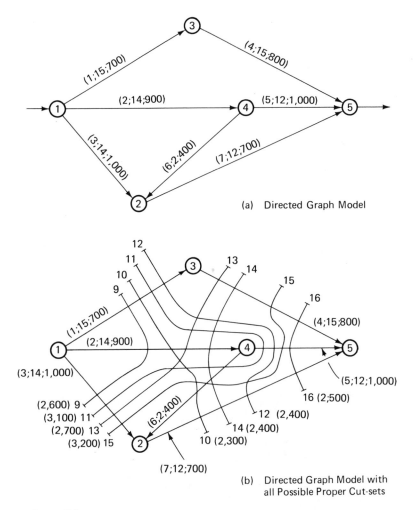

(a)   Directed Graph Model

(b)   Directed Graph Model with
      all Possible Proper Cut-sets

Figure 2.8

Linear Graph Model of Figure 2.5 with Flow in Road 6 Reversed

methods used the same graph, method one focuses on cut-sets and the other on paths.

2.5 LINEAR GRAPH TRANSFORMATIONS: PRIMAL-DUAL GRAPHS

In some cases, a problem formulation requires an analysis based on a graph property that is inconvenient for an analytical, logical or

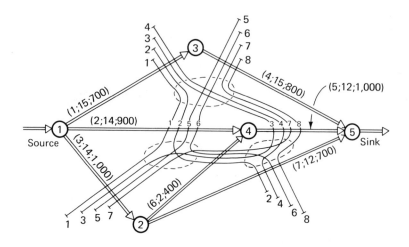

(a)   Road Network Cut-set Bundles

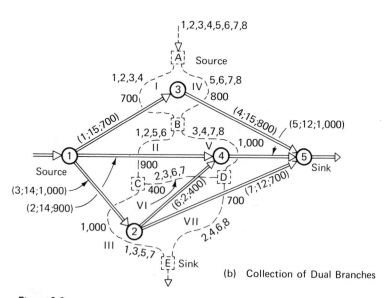

(b)   Collection of Dual Branches

Figure 2.9

Transformation of Primal Graph to Dual Graph

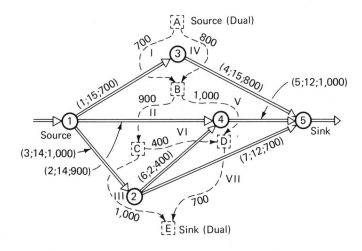

(c)   Road Network: Primal and Dual Graphs

Figure 2.9 Continued

computational reason. Unless another formulation is possible, it may be necessary to synthetically transform the original graph and graph property into an equivalent problem using a different graph model and graph property. In essence, the information contained in the linear graph is transformed into a more convenient and usable analysis format.

As an example, consider again the network flow problem and cut-set analysis approach. Suppose that the cut-sets of Figure 2.7 are redrawn as shown in Figure 2.9a in order to develop a concept for readily identifying and evaluating the cut-sets. In every case, the cut-sets mingle at focal points inside the graph. This suggests that the various cut-sets are obtained by linking these internal focal points in different ways and raises immediately a chain concept for the cut-set representation. Figure 2.9b shows this concept developed to the stage where the focal points of Figure 2.9a are replaced by square nodes and the bundles of cut-sets are replaced by dotted branches. All the cut-sets in the original or *primal* graph can be identified as different chains through the new dotted *dual* graph. Furthermore, each branch in the primal graph is intersected by a dotted branch in the dual. The graphical property of the primal is, in fact, the basis of the dual; and the cut-sets in the primal are transformed into chains in the dual. The transformation, however, is not complete unless the directions in the primal are translated to the dual. A simple sign convention can be easily derived from the cut-set principle. The following procedure can be followed:

1. Since there are two nodes in the dual graph that are outside the confines of the primal, designate one arbitrarily as the source node and the other as the sink node of the dual. It is generally more convenient to designate the top node as the source node in the dual graph.

2. Since any chain in the dual is a cut-set in the primal, if a branch in the primal crosses this cut-set from the *source* side to the *sink* side, then the corresponding branch in the dual should point along the direction of travel in the dual chain from the source in the dual to the sink in the dual.

3. If a branch in the primal crosses the cut-set from the *sink* side to the *source* side, the corresponding branch in the dual should point against the direction of travel in the dual chain, from the dual source to the dual sink.

For example, for the dual graph in Figure 2.9b, let node A be the dual source and node E be the sink. Since branch 1 crosses the chain (I-II-III) from the source side of the primal to the sink side of the primal, the corresponding branch I in the dual should be directed along the direction of travel from source A to E; i.e., from A to B in the chain (I-II-III). On the other hand, since branch 6 crosses the chain (IV-V-VI-III) from the sink side of the primal to the source side, the corresponding branch VI in the dual should be pointed against the direction of travel from A to E; i.e., from C to D in the chain (IV-V-VI-III).

Using the above sign convention, a unique direction is assigned to each branch in the dual, as shown in Figure 2.9c.

Network Analysis with Dual Graphs

The method for constructing the dual described above implies that the analysis of cut-sets in the primal can be replaced by a chain analysis in the dual. It is now necessary to identify the primal branch attribute of flow capacity with a corresponding dual branch attribute. The primal cut-set evaluation of capacity required the summation of individual branch capacities for each branch member of the cut-set. The corresponding evaluation in the dual requires the summation of the dual branch attributes along the relevant chain through the dual graph. In summing along a chain, it is convenient to think of the attributes as length. Hence, branch I of the dual has a length of 700, which corresponds to a capacity of 700 for branch 1 of the primal. Thus, for cut-set No. 3, the following one-to-one correspondence exists between the two methods of analysis:

PRIMAL (CUT-SET ANALYSIS)		DUAL (CHAIN ANALYSIS)	
Branch	*Capacity*	*Branch*	*Length*
1	700	I	700
5	1,000	V	1,000
6	0	VI	0
3	1,000	III	1,000
Total	2,700	Total	2,700

The primal problem of determining the minimum cut-set capacity now becomes the dual problem of determining the minimum length chain. The result of a complete chain analysis and its correspondence to the cut-set method is given in Table 2.2. The advantage of the chain method of analysis is that once the dual graph is constructed, all the possible chains going from the source to the sink can be easily and systematically enumerated.

Instead of enumerating all the possible chains, the minimum path can be found by selective addition of the dual branches. Thus, commencing at A, node B can be reached via branch I of length 700 or by branch IV of length 800; therefore, select branch I. Node C can be reached from node B via branch II or via branches V and VI for total lengths of 1,600 or 1,700; hence, select branch II. The remaining analysis portrayed in Figure 2.10a is left to the reader.

Table 2.2

CHAIN ANALYSIS

CHAIN	TOTAL LENGTH	CORRESPONDING PRIMAL CUT-SET
I-II-V	$700 + 900 + 1,000 = 2,600$	1
I-II-VI-VII	$700 + 900 + 400 + 700 = 2,700$	2
I-V-VI-III	$700 + 1,000 + 0 + 1,000 = 2,700$	3
I-V-VII	$700 + 1,000 + 700 = 2,400$	4
IV-II-III	$800 + 900 + 1,000 = 2,700$	5
IV-II-VI-VII	$800 + 900 + 400 + 700 = 2,800$	6
IV-V-VI-III	$800 + 1,000 + 0 + 1,000 = 2,800$	7
IV-V-VII	$800 + 1,000 + 700 = 2,500$	8

Maximum Flow Capacity = Minimum Chain Length = 2,400.

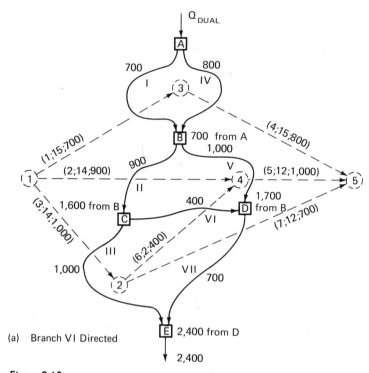

Figure 2.10

Minimum Chain Length in Dual Graph

The dual graph can also be used when the direction cannot be assigned *a priori* for the flow in some of the branches. Recall the original problem from Figure 2.5a. The direction of flow is known for all branches except branch 6. Therefore, draw the dual graph as shown in Figure 2.10b without an assigned direction to branch VI. The analysis proceeds the same as for Figure 2.10a until the computations are to be made for nodes *C* and *D*. Since the flow in branch VI may be either from *C* to *D* or from *D* to *C*, node *C* can be reached from node *B* via branch II for a length of 1,600 or via branches V and VI for a length of 2,100; hence, select branch II, which gives the minimum. Similarly, node *D* can be reached from node *B* via branch V for a length of 1,700 or via branches II and VI for a length of 2,000; hence, select branch V and the analysis proceeds as for Figure 2.10a; the minimum chain length is branches I, V, and VII, as before.

However, if road 6 must cross a river by means of a ferry that has a capacity of 50 vph, the length of branch VI would be 50 and the

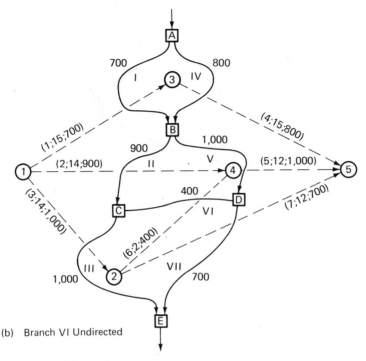

(b)   Branch VI Undirected

Figure 2.10 Continued

analysis would yield the following results. Since the flow in branch VI can be either C to D or D to C, node C can be reached from B via branch II for a length of 1,600 or via branches V and VI for a length of 1,750; hence, select branch II. Node D can be reached from B via branch V for a length of 1,700 or via branches II and VI for a length of 1,650; hence, select branches II and VI. Proceeding to node E, the minimum chain has a length of 2,350 and includes branches I, II, VI and VII. This indicates that the direction of branch VI must be from C to D and hence, the direction of flow in road 6 must be from node 2 to 4 in order to obtain the maximum flow capacity of the network. Thus, the dual graph analysis not only provides the maximum flow capacity of the network, but also provides a means of determining the direction of flow for branches on the critical chain for which flow directions cannot be assigned *a priori*.

Constructing the Dual Graph

Constructing the dual graph can be formalized into the following steps, as illustrated in Figure 2.11:

(b) Location of Dual Nodes

(e) Directed Dual Graph

[A] Source (Dual)

[E] Sink (Dual)

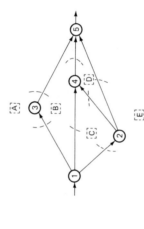

(c) Location of Dual Branches

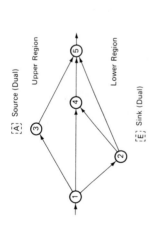

[A] Source (Dual)

Upper Region

Lower Region

[E] Sink (Dual)

(a) Location of Dual Source and Sink

(d) Undirected Dual Graph

Figure 2.11

Constructing the Dual Graph

58

1. Let the source and sink nodes of the primal subdivide the out-side region into an upper and a lower region; allocate the dual source to the upper region and the dual sink to the lower region;

2. Allocate a dual node inside each bounded region in the primal graph;

3. Cut each primal branch with a short line segment;

4. Connect each of the line sections to the dual node inside the same region;

5. Assign directions to the dual branches according to the sign convention established in the preceding section;

6. Transfer branch quantity from the primal to the dual.

Planar and Nonplanar Graphs

The above method of constructing the dual is applicable only to planar graphs. A planar graph is a graph in which the branches *can be*

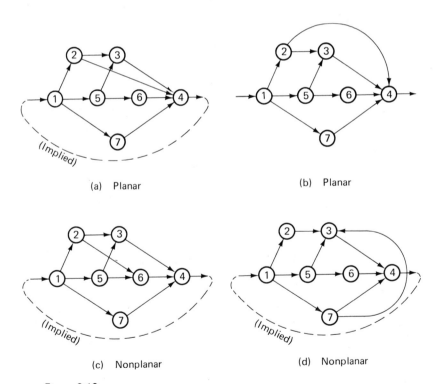

(a) Planar

(b) Planar

(c) Nonplanar

(d) Nonplanar

Figure 2.12

Planar and Nonplanar Graphs

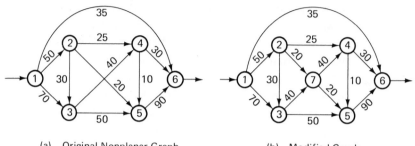

(a) Original Nonplanar Graph   (b) Modified Graph

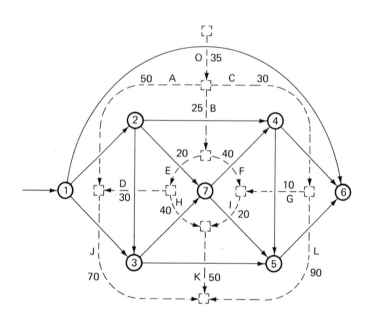

(c) Modified Graph: Primal and Dual

Figure 2.13

Dual of a Nonplanar Graph

drawn such that no two branches cross each other. Otherwise, the graph is called nonplanar.

For example, Figure 2.12a is a planar graph because it can be redrawn as in Figure 2.12b. According to the definition of a linear graph, these two graphs are identical because they express the same connectivity among the nodes and the branches. In constructing the dual of a planar graph, such as Figure 2.12a, it is always advisable to first redraw the branches so that no two cross each other. Both Figures 2.12c and 2.12d are nonplanar graphs.

By a simple modification of the preceding procedure, a dual graph can be developed for a nonplanar graph for the purpose of network analysis. For example, consider the nonplanar graph shown in Figure 2.13a. Suppose that this graph represents a water pipe network with the branches representing sections of pipes, the nodes representing junctions of pipes, and the numbers indicating the flow capacity in cubic feet per second (cfs.). The problem is to determine the maximum flow capacity from node 1 to node 6.

To construct the dual, a dummy node can be inserted at the intersections of branches 2-5 and 3-4, as shown in Figure 2.13b. Number this node 7. Since the new branches 2-7 and 7-5 are two sections of the same pipe, they must both be assigned the same direction as the branch 2-5 and have the same capacity of 20 cfs. Similarly, branches 2-7 and 7-4 must have the same direction as the replaced branch 3-4, and both have a capacity of 40 cfs. A dual can now be constructed for this modified graph using the preceding procedure. This step is illustrated in Figure 2.13c.

To use the dual in Figure 2.13c for network capacity analysis, it is necessary only to omit all the chains that pass through both branches *E* and *I* or both branches *H* and *F*. This is intuitively obvious, because such paths would correspond to cut-sets in the primal that cut the original branches 2-5 or 3-4 twice.

2.6 LINEAR GRAPH ANALYSIS: A NETWORK IMPROVEMENT PROBLEM

In order to illustrate the application of linear graph methods in network analysis, consider the problem faced by a state highway department as it prepares its request for funds for capital improvements for the network of roads between two cities.

General Statement of the Problem

The following are some of the more obvious issues that are posed by the problem of improving the road network system.

1. The network characteristics of the road network system,
   a. As currently existing,
   b. With recognized new routes;

2. The individual road characteristics,
   a. Length,
   b. Number of lanes, and hence, traffic capacity,
   c. Pavement status and, hence, level of service provided;

3. Improvement objective defined as,
   a. Improved network capacity,
   b. Improved level of service,
   c. Constant maintenance of current level of service;

4. Financial constraints defined as,
   a. Federal user funds for new road improvement construction,
   b. State taxation funds for maintenance and road improvement,
   c. Local taxation funds for road system maintenance;

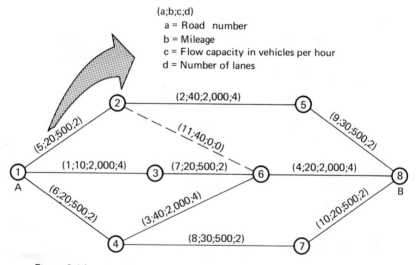

Figure 2.14

Road Network System Between Towns A and B

5. Road user's requirements,
   a. Simple demand traffic flows (special purpose),
   b. Complex network traffic flows,
   c. Social and political objectives.

In order to proceed with the linear graph analysis, consider the following specific case:

The problem concerns a transportation system between cities A and B passing through other intermediate towns, as shown in Figure 2.14. Towns and cities have been modeled as nodes and roads as branches. A feasible new road connecting towns 2 and 6 is shown as a dotted branch. Along each branch of the linear graph is a vector containing the relevant attributes of the road it represents: road identification number, mileage, flow capacity and number of lanes. The existing roads are either two or four lanes. A two-lane road has a capacity flow of 500 vehicles per hour in each direction and a four-lane road has a capacity flow of 2,000 vph in each direction, as illustrated in Figure 2.15.

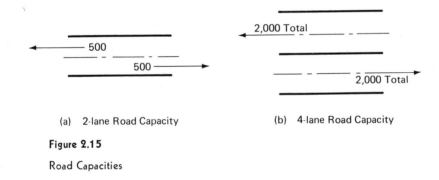

(a) 2-lane Road Capacity   (b) 4-lane Road Capacity

Figure 2.15

Road Capacities

Further investigation reveals that there is a demand for a flow of about 6,000 vph from city A to city B and that capacity improvements can be made by adding two lanes to existing two-lane roads to increase their capacity from 500 to 2,000 vph or by constructing a new road connecting towns 2 and 6. Road construction costs have been determined as follows:

1. The addition of two lanes to an existing two-lane road costs R per mile;
2. The construction of a new two-lane road costs $1.5R$ per mile;
3. The construction of a new four-lane road costs $2.3R$ per mile.

Assume that the intermediate towns between A and B do not generate sufficient traffic to affect traffic flow considerations between A and B.

Objectives and Investment Criteria

The highway department perceives its objective to be that of increasing the flow capacity of the road network between cities A and B such that it can accommodate 6,000 vph from A to B. The department has also established the following criteria for choosing among alternative methods of increasing the network capacity.

1. The alternative with the largest increase in network capacity per dollar of investment ($\Delta C/\$$) will be developed first.
2. If two or more alternatives have the same capacity increase per dollar of investment, the alternative that provides the most direct connection between A and B will be developed first.

The above objective and criteria are reasonably realistic, although in some instances political influences and regional developments affecting the intermediate towns may have to be considered. There may already have been some kind of influence exerted to cause road 3 between cities 4 and 6 to be built as a four-lane road, because as will be seen later, it is of no importance for maintaining or increasing the capacity of flow from A to B.

Analysis Procedure

Figure 2.16 is a linear graph model of the network flow showing all the routes going from city 1 (city A) to city 8 (city B). It is assumed that no traffic from city 1 to city 8 will take either the path consisting of roads 1, 7, 3, 8 and 10 or the path consisting of roads 1, 7, 11, 2, and 9 because of the unnecessarily long distances involved. This assumption allows the flow in each branch to be assigned a unique direction. If the distances were such that this assumption could not be made (e.g., if traffic might be expected to take one of these paths), then a procedure similar to the one described in Section 2.4 would have to be followed to determine the direction of flow on roads 3 and 11 if 11 were to be added.

The problem is simplified by considering only flow from city 1

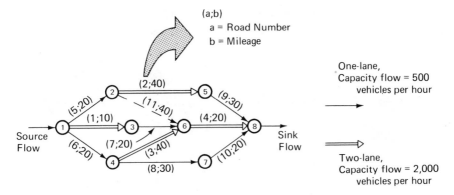

Figure 2.16

Network Flow Model

to city 8, since the reverse flow is identical in capacity. In Figure 2.16, the four-lane roads—e.g., those with two lanes in each direction—are represented by double-line branches.

A sequential procedure is followed. The roads that limit the flow capacity of the existing network are first identified. The feasible alternatives for increasing the capacity are outlined and ranked according to the established criteria. The best alternative is chosen and the limitation in the network with the added improvement is determined. This procedure is repeated until the total demand can be satisfied or no additional improvement is possible.

Figure 2.17a shows that the existing network has a maximum flow capacity of 1,500 vph and is limited by the roads included in two cut-sets. In order to expand the network capacity, both minimum cut-sets must be eliminated. Thus, the following feasible alternatives exist at this stage:

ALTERNATIVES	INCREASE IN CAPACITY (ΔC)	TOTAL COST	$\Delta C/\$$
1. Add two lanes to roads 5 and 9	1,500	\$ 50R	30/R
2. Add two lanes to road 7	1,500	\$ 20R	75/R
3. Add two lanes to road 6	1,500	\$ 20R	75/R
4. Add two lanes to road 5 and build four lanes at road 11	1,500	\$112R	13.4/R

Alternatives 2 and 3 result in the same capacity increase per dollar investment. But road 7 provides a more direct connection between

⟶   Capacity = 500 vehicles per hour
⟹   Capacity = 2,000 vehicles per hour

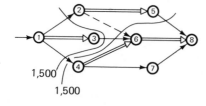

(a)   Existing Capacity

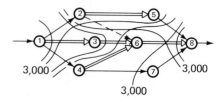

(b)   Capacity After Stage 1

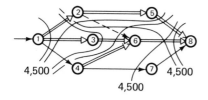

(c)   Capacity After Stage 2

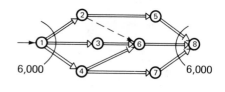

(d)   Final Network Capacity

Figure 2.17

Road Network Analysis

the two cities, 1 and 8. Thus, according to the second design criteria, the best choice at this stage is to add two lanes to road 7 at a cost of $20R$.

Figure 2.17b shows all the minimum cut-sets after the above improvement has been completed. The network capacity is now 3,000 vph. Using the same logic as before, in order to increase the network capacity, all these minimum cut-sets must be eliminated. Many feasible alternatives exist. But from Figure 2.17, since all the cut-sets pass through either branch 5 or branch 9, it is obvious that the two best alternatives are as follows:

ALTERNATIVES	INCREASE IN CAPACITY (ΔC)	COST ($)	$\Delta C/\$$
1. Add two lanes to roads 5 and 9	1,500	$50R$	$30/R$
2. Add two lanes to roads 6, 8, and 10	1,500	$70R$	$21.4/R$

Therefore, the best choice at this stage is alternative 1.

Figure 2.17c shows the minimum cut-sets after the second improvement has been completed. It can be easily shown that the best choice is to add two lanes to roads 6, 8, and 10 at the cost of $70R$, resulting in an increase in network capacity of 1,500 vph and, hence, providing a total capacity of 6,000 vph, as shown in Figure 2.17d.

Thus, the department has developed a cost-capacity relationship for the road network, as indicated by Table 2.3. Therefore, in order to provide capacity for 6,000 vph, the department should request a capital appropriation of $140R$ in order to add two lanes to roads 5, 6, 7, 8, 9, and 10. If the department receives a capital appropriation below its $140R$ request, Table 2.3 indicates how the money should be invested; e.g., if an

Table 2.3

COST CAPACITY RELATIONSHIPS FOR ROAD NETWORK

CAPACITY	TOTAL COST OF CAPITAL IMPROVEMENT	NO. LANES	ROAD
1,500	0		
3,000	20R	2	7
4,500	70R	2	5, 7, 9
6,000	140R	2	5, 6, 7, 8, 9, 10

appropriation of only $70R$ is received, two lanes should be added to roads 5, 7, and 9. This addition increases the capacity to only 4,500 vph, and an attempt must be made to secure the additional $70R$ if the total demand is to be satisfied.

Another way of interpreting these results is that Table 2.3 shows the minimum cost of providing different network capacities. Thus, the analysis would have been the same if the department had stated that it wanted to minimize the cost of providing a network capacity of 6,000 vph between cities 1 and 8.

2.7   LINEAR GRAPH TOPOLOGICAL MATRICES

The many graphical structural properties provide the opportunity for storing information about specific linear graphs. For example, a path can be identified with the nodes and branches that are traversed along the path. A convenient way of describing graph structure is through the use of matrices. A matrix consisting of a rectangular array of elements has a unique capacity for fulfilling this function.

In some cases, the ultimate usefulness of the linear graph method of analysis depends on the ability to describe the graphs in some numerical format so that complex computations may be performed logically, either by hand or with the help of an electronic computer.

In addition, matrix algebra notation represents a system of grouping numerical quantities in such a way that a single symbol suffices to denote the whole group. In this way, an algebraic form of shorthand can be developed. A review of matrix algebra is provided in Appendix A.

Since matrices are commonly considered as two-dimensional arrays with rows and columns, they can be used to model binary relationships. Thus, a path matrix could be developed to sort the various nodes or branches into the different paths existing in the linear graph. A row could be provided and identified for each node (or branch) and a column provided for each unique path. An entry in the matrix could then signify whether or not a node (or branch) is a component of the graph structural property considered in the path. In this way, a node-path matrix or a branch-path matrix could be developed. Finally, a sign convention can be introduced to capture the additional information portrayed in directed graphs.

The development of a linear graph topological matrix therefore requires:

1. Selecting the structural property attribute to be described and the graph component attribute (node or branch) against which the structural property is to be mapped;

2. Allocating the two attributes to the matrix rows and columns;

3. Selecting the mapping rule that distinguishes whether the relationship exists (i.e., true ≡ 1) or not (i.e., false ≡ 0);

4. Selecting the sign convention to be followed for directed graphs and directed structural properties.

In the following sections, several topological matrices will be developed to illustrate the general approach to the matrix algebra modeling of graph structure.

The Branch-Node Incidence Matrix A

The fundamental property of a linear graph is the connectivity of its branches and nodes. A linear graph is completely specified once the incidence of each branch on each node is known. This is conveniently done by specifying a matrix $\bar{\mathbf{A}}$ called the augmented branch-node incidence matrix, consisting of b rows (one per branch) and n columns (one per node). For undirected graphs, the typical element $\bar{a}_{ij} = (1,0)$ if the ith branch (is, is not) incident on the jth node (see Figure 2.18a). For

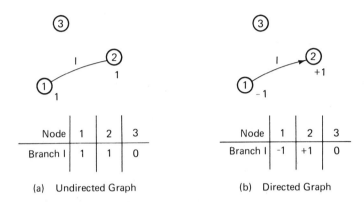

(a)   Undirected Graph   (b)   Directed Graph

Figure 2.18

Branch Node Incidence

directed graphs, the typical element $\bar{a}_{ij} = (+1, -1, 0)$ if the ith branch is (positively, negatively, not) incident on the jth node (see Figure 2.18b).

Each row corresponds to a particular branch that connects two nodes, and has two nonzero elements that for directed graphs are one $+1$ and one -1. Each column corresponds to a particular node, and the number of nonzero elements in it indicates the number of branches that are incident on the node. In addition, by summing independently the number of the positive and negative entries in a column, the number of originating and terminating branches on the node can be determined. The $2b$ nonzero entries in the augmented branch-node incidence matrix completely specify an orientated linear graph if the branch and node ordering within the rows and columns are known.

As an illustration, consider the directed linear graph shown in Figure 2.19. The following table can be identified:

NODES

		n_0	n_1	n_2	n_3
	b_1	-1	$+1$	0	0
	b_2	-1	0	$+1$	0
BRANCHES	b_3	0	-1	$+1$	0
	b_4	0	-1	0	$+1$
	b_5	0	0	-1	$+1$

and in matrix algebra form

$$\underset{(5\times4)}{\bar{\mathbf{A}}} = \begin{bmatrix} -1 & +1 & 0 & 0 \\ -1 & 0 & +1 & 0 \\ 0 & -1 & +1 & 0 \\ 0 & -1 & 0 & +1 \\ 0 & 0 & -1 & +1 \end{bmatrix} \tag{2.6}$$

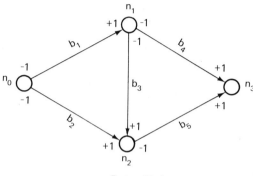

n_0 = Datum Node

Figure 2.19

Notation for Branch-node Incidence Matrix

where the rows and columns have been identified and labeled. Once the row and column labels have been prescribed, it is no longer essential to include them as descriptors on the matrix itself.

It is easy to see that the $\bar{\mathbf{A}}$ matrix contains redundant information. Since each row contains exactly one $+1$ coefficient and one -1 coefficient, then independent of their column locations within the row, the sum of all coefficients in the row is zero. That is to say, for a connected graph, the columns are one-fold linearly dependent and one nodal column (i.e., the datum node) can be suppressed without loss of information.

The reduced matrix is called the branch-node incidence matrix $\mathbf{A}$ and has b rows and $n - 1$ columns for a connected graph. The suppressed nodal (datum) column can be chosen at will from any node in the connected graph, but is usually chosen for convenience as the first or last within the nodal sequence of $\bar{\mathbf{A}}$. Practical examples of datum nodes include the starting node of a CPM network, the source node of a flow network, and the beginning node of a decision tree.

Using n_0 as the datum node, the branch-node incidence matrix $\mathbf{A}$

becomes

$$
\underset{(5\times3)}{\mathbf{A}} = \begin{bmatrix} +1 & 0 & 0 \\ 0 & +1 & 0 \\ -1 & +1 & 0 \\ -1 & 0 & +1 \\ 0 & -1 & +1 \end{bmatrix} \tag{2.7}
$$

In analyzing the road network capacity problem, it was necessary to identify the various proper cut-sets and highway branches of the linear graph model. A cut-set matrix, **D**, may be derived directly from the entries in Table 2.1 for the directed graph in Figure 2.7 as follows:

$$
\underset{(7\times8)}{\mathbf{D}} = \begin{bmatrix} 1 & 1 & 1 & 1 & 0 & 0 &) & 0 \\ 1 & 1 & 0 & 0 & 1 & 1 & 0 & 0 \\ 1 & 0 & 1 & 0 & 1 & 0 & 1 & 0 \\ 0 & 0 & 0 & 0 & 1 & 1 & 1 & 1 \\ 0 & 0 & 1 & 1 & 0 & 0 & 1 & 1 \\ 0 & 1 & -1 & 0 & 0 & 1 & -1 & 0 \\ 0 & 1 & 0 & 1 & 0 & 1 & 0 & 1 \end{bmatrix} \tag{2.8}
$$

where $d_{ij} = -1$ if the ith branch is included in the jth cut-set and flows from the sink side to the source side; $d_{ij} = 0$ if the ith branch is not included in the jth cut-set; and $d_{ij} = +1$ if the ith branch is included in the jth cut-set and flows from the source side to the sink side. The above **D** matrix adequately describes the components of all the possible proper cut-sets of the directed graph in Figure 2.7.

For undirected graphs, the d_{ij} elements of the matrix may simply be either 1 or 0, denoting whether the ith branch is or is not included in the jth cut set.

Other topological matrices may be constructed to describe other graphical properties in the same manner (Busacker and Saaty, 1965; Marshall, 1971). The needs for such matrices depend on the method of analysis and the graphical properties of the network.

2.8 SUMMARY

Many engineering systems can be modeled and analyzed using linear graph models and concepts. Linear graph analysis requires that the problem be expressed in graphical form, that the problem be identified with a graphical property, and that an analysis procedure be developed that focuses on this specific graphical property. The problem can often be identified as a path problem or a cut-set problem.

If a problem formulation requires an analysis that is inconvenient, it may be possible to use dual graph concepts to transform the original graph model and graphical property into an equivalent problem using a different graph model and graphical property.

Topological matrices provide a numerical format in which to describe linear graphs and graphical properties. The use of linear graph modeling and topological matrices in analyzing engineering systems are discussed further in Chapter 3.

2.9 PROBLEMS

P2.1. For each of the following situations identify system components, linear graph structure, and characteristics. To what extent can linear graphs aid in representing and understanding the situation.

1. A school bus route
2. A river system
3. A variety of road and highway intersections
4. A plant layout
5. A city water distribution system
6. A bridge span

P2.2. Identify a problem that might be associated with each of the situations in P2.1. Develop a linear graph and describe a linear graph procedure that can be used to solve the problem.

P2.3. Figure P2.3 represents the portion of a street network in a city. The direction of travel that is permitted on each street is indicated by the arrows in the figure. Also, the capacity of each street segment is shown in vehicles per hour.

1. Currently, the street segment between intersections C and D is two-way with a capacity of 500 vehicles per hour in each direction. A plan is being considered to convert this segment to one-way operation with a capacity of 1,500 vehicles per hour in order to increase the flow of traffic. Would such a plan increase the flow of traffic in the network between intersections A and G? If so, which direction should the one way street be? Justify your answer.

2. Based on your decision in (1), what is the maximum flow capacity between intersections A and G?

3. A traffic study reveals that the parking on street segments II, IV, and IX might be banned to increase the street capacity. For each street segment, the parking ban will increase the capacity by 400 vehicles per hour. Because of a shortage of parking in the area, the city council has consented to ban parking from only one segment. Which one should be chosen to increase the flow capacity of the street network between intersections A and G?

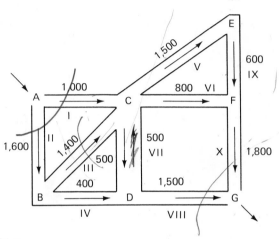

Figure P2.3

Street Network

P2.4. Develop a flow process linear graph for manufacturing precast concrete wall panels using metal molding forms and steam curing. The panels are reinforced with steel mesh and contain utility ducting as well as both window and door frame inserts. Use the following mode symbols and attributes:

- ● Operation (description)
- ➤ Transport (distance, mode)
- ▲ Storage (time period, quantity)
- ■ Inspection (detail, quality)

P2.5. Identify the components and component characteristics for the following systems in a high-rise building. Develop linear graph models for each system and explain how linear graph analysis procedures can aid in the design of each system:

1. An electrical network
2. A heating and ventilation system
3. A hot and cold water distribution system
4. A waste collection system
5. A transportation system
6. A structural system
7. A telephone system

P2.6. A city plans to expand its water supply by developing a well field northeast of town, as shown in Figure P2.6. The water must be

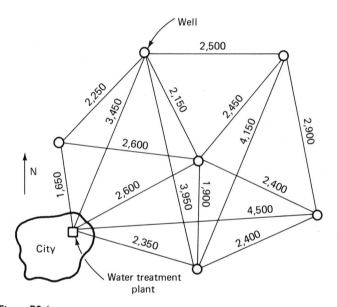

Figure P2.6

pumped from the wells to the water treatment plant before distribution. The distances in feet are shown in Figure P2.6.

1. In what pattern should the wells be connected in order to use the minimum length of pipe?
2. In (1) the minimum length of pipe was used as a criterion for defining the best solution. What might be another criterion to use for this problem? Can the same analysis procedure be used for both criteria?

P2.7. Two bridges carry traffic across a river from an adjacent city road network, as shown in Figure P2.7. The bridges have traffic

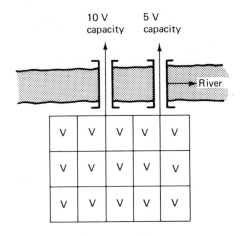

Figure P2.7

capacities of $10V$ and $5V$, V being the steady state traffic volume generated by each city block.

How would you approach the problem of determining the steady state traffic flow pattern using linear graph concepts?

P2.8. The way in which a problem is formulated will often control which analysis procedure can be used. The linear graph formulation used for the problem in Section 2.6 resulted in the use of a cut-set analysis procedure. Rework this problem using the concept of a dual graph. What are some possible advantages of attempting several different formulations of the problem using different graphical properties with respect to:

1. The analysis procedure that can be used?
2. The amount and kinds of data required?
3. The kinds and detail of answers that can be obtained?

3

MATHEMATICAL MODELING
OF ENGINEERING SYSTEMS

The Parque Aukembi exhibition center in Sao Paulo, Brazil, features a space-frame roof that is 850 ft. by 850 ft. The space-frame has a grid of 11 ft. square and a depth of 7 ft. 9 in. The space-frame transmits the roof loads to the vertical columns. (From ASCE Preprint 1391 by Cedric Marsh and Peter K. Kneen, Courtesy of the Civil Engineering Magazine, American Society of Civil Engineers)

3.1 INTRODUCTION

Models are extensively used by engineers: as aids in the description, analysis, and design phases of problem solving; to facilitate the communication of ideas to others; and as a means of storing information for future reference and use.

A model is a representation of some object or condition that exists, or of an issue under consideration. It contains information expressed in certain specific forms about the object it models and requires interpretation according to certain predefined rules. Often the modeling medium is the symbolism of mathematics and requires the logic and understanding of mathematics for its interpretation. In many cases, mathematical models require the support of both descriptive and schematic models to define the nature of the model and to aid in interpreting the symbols and variables used in the model.

A mathematical model can thus be considered to be made up of four basic component models, namely:

1. A descriptive model of the problem and purpose;

2. A symbolic definition model describing all the model symbols, variables, parameters, and constants used in the mathematical model itself;

3. A schematic model illustrating the interpretation of the symbols and interrelationships among the system components;

4. The mathematical model itself.

As indicated in Figure 3.1, the first three component models define

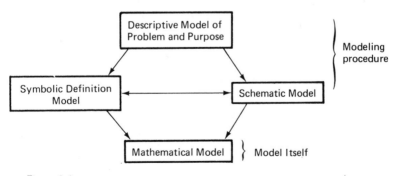

Figure 3.1

Basic Components of a Mathematical Model

the manner in which the final mathematical model (component 4) is to be interpreted.

In the mathematical modeling of engineering systems, each component of the system may require modeling components as defined above and as illustrated in Figure 3.1.

3.2 THE FUNDAMENTAL USE OF MODELS

As an illustration of the modeling process and the various kinds of models used in problem-solving, consider the following example:

Model 1: Descriptive Model of Problem and Purpose

> A straight uniform bar of specified length is to be made from a light elastic material. The bar is to hang vertically from a support and carry a specific axial load at its free bottom end within a specific deflection range.

This model is a statement of the modeler's view and evaluation of the problem that exists for him in reality. It formulates the model purpose and system constraints; its definition is a critical step since it provides the requirements and objectives for the final model.

Model 2: The Symbolic Definition Model

Let A be the connection point of the bar to the support

B be the bottom end point of the bar

L be the length of the bar in inches

A_x be the cross-sectional area of the bar in square inches

E be the Young's modulus of the material in pounds per square inch

P be the axial force along the bar member in pounds

σ be the bar member stress in pounds per square inch

ϵ be the bar member strain

ΔL be the bar member elongation in inches

P'_B, P'_A be the applied axial load at end B, and the reaction load at A, respectively, in pounds

u'_B, u'_A be the displacements of the points B and A, respectively, in inches

Δ be the maximum permissible movement of the applied load at B in inches.

Model 2 is the interface between the modeler's knowledge of the real problem and the requirements of a specific technological model. These symbols, in fact, represent the system components and their attributes that are relevant to the problem.

Model 3: The Schematic Model

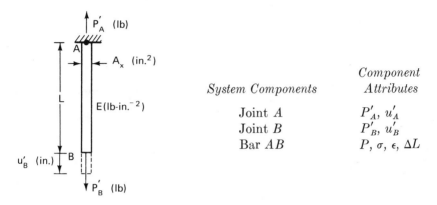

System Components	Component Attributes
Joint A	$P'_A,\ u'_A$
Joint B	$P'_B,\ u'_B$
Bar AB	$P,\ \sigma,\ \epsilon,\ \Delta L$

Model 4: Mathematical Model of the Bar Behavior

Bar assumed weightless

$P = $ constant

$\sigma = P/A_x$

$\epsilon = \Delta L/L = \sigma/E = P/A_x E$

mathematical logic and derivation; i.e., program segment for deriving mathematical model

$$\therefore \Delta L = \frac{PL}{A_x E} \quad \dots\dots\dots\dots \quad \text{mathematical model}$$

Model 4 is a mathematical model for the technological description of the bar behavior. Its selection depends on the level of technology available and known to the modeler and the relevant depth considered necessary to meet the problem requirements. The mathematical model identifies the five parameters that describe the system response and structure of the bar. If any four parameters are known, the fifth can be found directly from the model. It can be used to study the bar response to the design problem model described later.

Model 5: Mathematical Model of the System Component Interaction

$$\Delta L = u'_B - u'_A$$

$$u'_A = 0$$

$$P = P'_B$$

$$P'_A = P'_B$$

Model 5 defines the system structure relating the component bar parameters and the node parameters. It normally requires a unique creative effort on the modeler's part, since a specific technological formulation for the problem may not exist.

Model 6: Mathematical Model of the System Response

From model 4 $\quad \Delta L = \dfrac{PL}{A_xE} \quad$ where P'_B is the system input

From model 5 $\quad \Delta L = u'_B \qquad \left(\dfrac{L}{A_xE}\right)$ is the system mode of response

$$P = P'_B \qquad \qquad u'_B \text{ is the system response and system output}$$

$$\therefore\ u'_B = \frac{P'_BL}{A_xE}$$

Model 6 is a statement indicating how the modeler intends to *read* the system. It focuses on the input-output description of the system and represents the solution process for the systems model.

Model 7: Mathematical Model of the Design Problem

$$u'_B = \frac{P'_BL}{A_xE} \leq \Delta$$

$$\therefore\ A_xE \geq \frac{P'_BL}{\Delta} = \text{a constant}$$

Model 7 is a mathematical model exposing the design variables A_x and E. The modeler can now establish a design criterion to test whether a particular design is suitable or not. Since the design criterion is bounded on one side only, the designer must establish his own value system for departures of A_xE from the constant P'_BL/Δ.

The total mathematical model for the system design problem is summarized in Figure 3.2. The seven model components portray in various ways such things as the embodiment of model purpose, scope of investigation, reality constraints on conditions and choices, and the modeler's bias and knowledge of technology. They do not, of course, model the design process and selection of specific attributes for the model components.

The model itself can be checked for internal consistency; for level of accuracy relevant to the real-world problem, with problem identification, needs, and resources; and for the impact of the modeler's bias.

Model 1, for example, prescribes that the structure be a bar (not a truss or combination of structural elements), that it be uniform, straight,

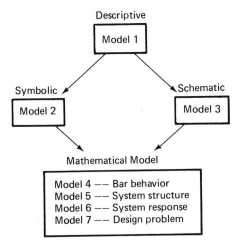

Figure 3.2

Mathematical Model of System Design Problem

and be made from a light elastic material. The inclusion of all (or any one) of these descriptions severely limits the design choices potentially available and, in effect, plays a dominant role in the specific forms of models 2 to 7.

The requirement for a light material (the validity of which should be considered) tentatively supports the assumption of a weightless bar in model 4 and the vague selection criterion imposed in model 7. Existing technology permits model 4 to be updated to include the effect of the self-weight of the bar. Whether this is done or not reflects a decision made by the modeler. If the material selection aspect of the problem is to be elaborated, at least models 1 and 7 are affected and require replacement by new or updated versions.

Similarly, an investigation can be directed to the necessity for the imposition of constraints such as straightness and uniformity on the bar. If the load is applied dynamically or suddenly, model 4 may not be accurate or suitable, in which case, models 1, 4, 6, and 7 must be changed.

Suppose that the cost of the structure and the time to manufacture and erect the structure enter into consideration and affect the model's purpose. The requirement for internal consistency in the models demands that if costs enter into model 1, a consideration of costs must appear in at least one of the following models. The specific form of its inclusion will expose the modeling and design rationale of the modeler, which can then be examined and critiqued.

The development of useful and valid models requires a high level of creative ability, technical and professional knowledge, and an understanding of the forces, constraints, and values at work in the environment in which the problem exists.

3.3 A COMPOUND BAR SYSTEM MODEL

If a single equation can be considered as a model of a condition or component, then a set of equations intuitively represents a model of a system. In fact, many system models take the form of a set of equations, either algebraic, functional, differential, or integral.

In some cases, each individual equation introduces a new condition initiated by the consideration of another component in the system model. In other cases, each equation introduces a new constraint on the relationships among the set of system variables.

Great insight into mathematical models can be gained if the various equations are grouped into sets in which each set of equations focuses on a specific type of system constraint; i.e., component behavior, system structure, system equilibrium.

As an example, consider a compound bar system in structural design.

Statement of the Problem

The compound bar system shown in Figure 3.3 consists of four bar components, 1, 2, 3, and 4, connected to three short horizontal members, A, B, and C, in which member A is a fixed support. The construction is such that these horizontal members can be considered as rigid cross pieces so that the vertical bar members are constrained to axial deformations only; i.e., the horizontal members remain horizontal and can be treated as individual joints in the structure.

The system can be loaded with joint forces P'_B and P'_C at B and C, developing the reaction force P'_A. The joint forces P'_B and P'_C can be considered as system inputs.

As a result of the applied system loading, the bars will be loaded, producing axial forces P_1, P_2, $\ldots$, P_4, and the bars will deform by the amounts u_1, u_2, $\ldots$, u_4. This internal loading and deformation of the system will cause joints B and C to displace amounts u'_B and u'_C. Since joint A is a fixed support, the displacement u'_A is zero. The joint dis-

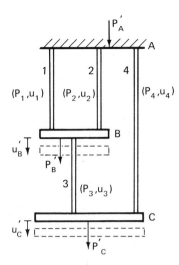

Figure 3.3

Compound Bar System

placements u'_B and u'_C can be considered as measurements of the system response (i.e., output) to the system loading input.

The problem is to develop a mathematical behavior model that can be used to analyze the system response under various combinations of loading at B and C. This is realized if a set of equations can be developed to express the system response parameters u'_B and u'_C as functions of the system input parameters P'_B and P'_C.

System Constraints

The following constraints exist in the structure:

1. Component behavior—The response of a bar member when subjected to an axial load is governed by a physical law, commonly referred to as an equation of state;

2. System compatibility—Geometric relationships exist between the elongation of the bar members and the physical displacements of the joints;

3. System equilibrium—The components are in stable equilibrium under static loading.

Component Behavior

It has already been shown that the following relationship exists between the axial force P_i acting on a bar and the corresponding elongation u_i:

$$P_i = \left(\frac{E_i A_i}{L_i}\right) u_i = k_i u_i \tag{3.1}$$

In structural engineering, k_i is called the stiffness coefficient of bar i and is expressed in units of bar force per unit displacement (lb/in.). The stiffness coefficient k_i thus reflects the "stiffness" of bar i and is a function of the material, the cross-sectional area and its length.

One such equation can be used to describe the behavior response of each of the four members in the given structure. Thus,

$$
\begin{aligned}
P_1 &= k_1 u_1 \\
P_2 &= k_2 u_2 \\
P_3 &= k_3 u_3 \\
P_4 &= k_4 u_4
\end{aligned}
\tag{3.2}
$$

In matrix notation, Equations 3.2 become:

$$
\begin{bmatrix} P_1 \\ P_2 \\ P_3 \\ P_4 \end{bmatrix}
=
\begin{bmatrix}
k_1 & 0 & 0 & 0 \\
0 & k_2 & 0 & 0 \\
0 & 0 & k_3 & 0 \\
0 & 0 & 0 & k_4
\end{bmatrix}
\begin{bmatrix} u_1 \\ u_2 \\ u_3 \\ u_4 \end{bmatrix}
\tag{3.3}
$$

i.e.,
$$\mathbf{P}_{(4\times 1)} = \mathbf{K}_{(4\times 4)} \mathbf{u}_{(4\times 1)} \tag{3.4}$$

where

$\mathbf{P}_{(4\times 1)}$ is the column matrix of component axial forces;

$\mathbf{u}_{(4\times 1)}$ is the column matrix of component axial deformations;

$\mathbf{K}_{(4\times 4)}$ is a square diagonal matrix of component stiffnesses. Its diagonal form indicates (and establishes through matrix multiplication with $\mathbf{u}_{(4\times 1)}$) the independence of the four equations.

System Compatibility

Connecting the bars together to form the system structure requires the development of mathematical expressions to enforce the geometric compatibility between member elongations and the joint displacements.

System compatibility is insured if the individual members continue to be connected to the joints whatever the joint displacements. This implies that the members must elongate just enough to suit the joint displacements. Therefore:

$$u_1 = -u'_A + u'_B$$
$$u_2 = -u'_A + u'_B$$
$$u_3 = -u'_B + u'_C$$
$$u_4 = -u'_A + u'_C$$

(3.5)

Equations 3.5 state that member elongations depend on relative joint displacements.

In matrix notation,

$$\begin{bmatrix} u_1 \\ u_2 \\ u_3 \\ u_4 \end{bmatrix} = \begin{bmatrix} -1 & +1 & 0 \\ -1 & +1 & 0 \\ 0 & -1 & +1 \\ -1 & 0 & +1 \end{bmatrix} \begin{bmatrix} u'_A \\ u'_B \\ u'_C \end{bmatrix}$$

(3.6)

i.e.,
$$\mathbf{u}_{(4\times1)} = \bar{\mathbf{A}}_{(4\times3)}\mathbf{u}'_{(3\times1)}$$

(3.7)

where

$\mathbf{u}'_{(3\times1)}$ is the column matrix of joint displacements;

$\bar{\mathbf{A}}_{(4\times3)}$ is a rectangular matrix of coefficients that yield relative displacements on matrix multiplication with $\mathbf{u}'_{(3\times1)}$.

System Equilibrium

Assuming that the applied loads P'_B and P'_C are static, the system is in equilibrium if the algebraic sum of the forces acting on any joint in the structure is equal to zero. Any direction (up or down) can be assumed to be positive since a reversal of sign does not affect the equation, but a positive sense must be maintained during the formation of the equations. Usually the same positive direction is used for all joints in the structure.

Figure 3.4 illustrates the forces acting on the present structure. Bar axial forces are shown as tensile (i.e., positive) forces, and joint forces as downward (i.e., positive). The following equilibrium equations can be derived:

At joint A $P'_A + P_1 + P_2 + P_4 = 0$

At joint B $P'_B - P_1 - P_2 + P_3 = 0$ (3.8)

At joint C $P'_C - P_3 - P_4 = 0$

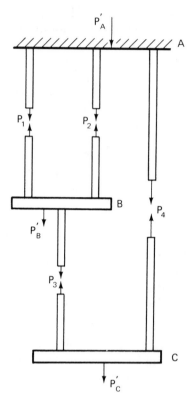

Figure 3.4

Member Forces and Joint Loads

Thus, rearranging terms to separate the applied joint loads (cause) from the bar component member loads (effect) results in:

$$P'_A = -P_1 - P_2 \qquad - P_4$$
$$P'_B = +P_1 + P_2 - P_3 \qquad (3.9)$$
$$P'_C = \qquad\qquad + P_3 + P_4$$

In matrix notation

$$
\begin{bmatrix} P'_A \\ P'_B \\ P'_C \end{bmatrix} =
\begin{bmatrix} -1 & -1 & 0 & -1 \\ +1 & +1 & -1 & 0 \\ 0 & 0 & +1 & +1 \end{bmatrix}
\begin{bmatrix} P_1 \\ P_2 \\ P_3 \\ P_4 \end{bmatrix}
\qquad (3.10)
$$

By comparing the coefficient matrix of Equation 3.10 with the coefficient matrix $\bar{\mathbf{A}}$ in Equation 3.7,

$$\mathbf{P}'_{(3\times1)} = \bar{\mathbf{A}}^T_{(3\times4)}\mathbf{P}_{(4\times1)} \tag{3.11}$$

in which $\mathbf{P}'$ is the column matrix of joint loads and $\bar{\mathbf{A}}^T$ is the transpose of $\bar{\mathbf{A}}$, i.e., the rows and columns have been interchanged.

System Model

A complete mathematical model of the compound bar system is obtained by grouping Equations 3.4, 3.7, and 3.11 together as follows:

4 equations of state

$$\mathbf{P}_{(4\times1)} = \mathbf{K}_{(4\times4)}\mathbf{u}_{(4\times1)} \tag{3.4}$$

4 compatibility equations

$$\mathbf{u}_{(4\times1)} = \bar{\mathbf{A}}_{(4\times3)}\mathbf{u}'_{(3\times1)} \tag{3.7}$$

3 equilibrium equations

$$\mathbf{P}'_{(3\times1)} = \bar{\mathbf{A}}^T_{(3\times4)}\mathbf{P}_{(4\times1)} \tag{3.11}$$

The eleven equations contain the fourteen system variables P_1, P_2, P_3, P_4, u_1, u_2, u_3, u_4, P'_A, P'_B, P'_C, u'_A, u'_B, and u'_C. For any solution, three variables must be specified. In this particular example, $u'_A = 0$ and P'_B and P'_C are known from the given loading condition.

Furthermore, by performing some simple substitutions, a mathematical model can be developed to express the relationship between the system response vector $\mathbf{u}'$ and the system input vector $\mathbf{P}'$. Thus, by substituting Equation 3.4 into Equation 3.11, the following results:

$$\mathbf{P}' = \bar{\mathbf{A}}^T\mathbf{K}\mathbf{u}$$

Then, substituting Equation 3.7 into this equation yields

$$\mathbf{P}' = (\bar{\mathbf{A}}^T\mathbf{K}\bar{\mathbf{A}})\mathbf{u}' \tag{3.12}$$

which is the desired mathematical behavior model. The three simultaneous equations of Equation 3.12 are the joint equations for equilibrium in terms of the joint displacements that contain implicitly the conditions built in by the equations of state, Equation 3.4, and bar member compatibility, Equation 3.7.

In long-hand form, Equation 3.12 becomes:

$$\begin{bmatrix} P'_A \\ P'_B \\ P'_C \end{bmatrix} = \begin{bmatrix} -1 & -1 & 0 & -1 \\ +1 & +1 & -1 & 0 \\ 0 & 0 & +1 & +1 \end{bmatrix} \begin{bmatrix} k_1 & 0 & 0 & 0 \\ 0 & k_2 & 0 & 0 \\ 0 & 0 & k_3 & 0 \\ 0 & 0 & 0 & k_4 \end{bmatrix} \begin{bmatrix} -1 & +1 & 0 \\ -1 & +1 & 0 \\ 0 & -1 & +1 \\ -1 & 0 & +1 \end{bmatrix} \begin{bmatrix} u'_A \\ u'_B \\ u'_C \end{bmatrix}$$

and using matrix multiplication

$$
\begin{bmatrix} P'_A \\ P'_B \\ P'_C \end{bmatrix} = \begin{bmatrix} -k_1 & -k_2 & 0 & -k_4 \\ k_1 & k_2 & -k_3 & 0 \\ 0 & 0 & k_3 & k_4 \end{bmatrix} \begin{bmatrix} -1 & +1 & 0 \\ -1 & +1 & 0 \\ 0 & -1 & +1 \\ -1 & 0 & +1 \end{bmatrix} \begin{bmatrix} u'_A \\ u'_B \\ u'_C \end{bmatrix}
$$

there results:

$$
\begin{bmatrix} P'_A \\ P'_B \\ P'_C \end{bmatrix} = \begin{bmatrix} (k_1 + k_2 + k_4) & (-k_1 - k_2) & (-k_4) \\ (-k_1 - k_2) & (k_1 + k_2 + k_3) & (-k_3) \\ (-k_4) & (-k_3) & (k_3 + k_4) \end{bmatrix} \begin{bmatrix} u'_A \\ u'_B \\ u'_C \end{bmatrix} \quad (3.13)
$$

Equation 3.13 in effect contains the following three algebraic equations:

$$
P'_A = (k_1 + k_2 + k_4)u'_A - (k_1 + k_2)u'_B - k_4 u'_C \quad (3.14)
$$

$$
P'_B = -(k_1 + k_2)u'_A + (k_1 + k_2 + k_3)u'_B - k_3 u'_C \quad (3.15)
$$

$$
P'_C = -k_4 u'_A - k_3 u'_B + (k_3 + k_4)u'_C \quad (3.16)
$$

Since $u'_A = 0$, these equations may be simplified as follows:

$$
P'_A = -(k_1 + k_2)u'_B - k_4 u'_C \quad (3.17)
$$

$$
P'_B = (k_1 + k_2 + k_3)u'_B - k_3 u'_C \quad (3.18)
$$

$$
P'_C = -k_3 u'_B + (k_3 + k_4)u'_C \quad (3.19)
$$

Thus, given P'_B and P'_C, u'_B and u'_C can be determined by solving Equations 3.18 and 3.19. The reaction joint load P'_A can, in turn, be computed from Equation 3.17. In matrix notation, Equations 3.18 and 3.19 may be written as

$$
\begin{bmatrix} P'_B \\ P'_C \end{bmatrix} = \begin{bmatrix} (k_1 + k_2 + k_3) & -k_3 \\ -k_3 & (k_3 + k_4) \end{bmatrix} \begin{bmatrix} u'_B \\ u'_C \end{bmatrix} \quad (3.20)
$$

and the solution is

$$
\begin{bmatrix} u'_B \\ u'_C \end{bmatrix} = \begin{bmatrix} (k_1 + k_2 + k_3) & -k_3 \\ -k_3 & (k_3 + k_4) \end{bmatrix}^{-1} \begin{bmatrix} P'_B \\ P'_C \end{bmatrix} \quad (3.21)
$$

which represents the final input-output response model. This inverse matrix embodies all three types of system constraints in the problem: component behavior, system compatibility, and system equilibrium, and reflects the materials used for the bar members, the geometry and equilibrium of the structure. Thus, for a given design of the structure, this inverse matrix remains constant and the mathematical model in Equation 3.21 can be used to solve for the joint displacements u'_B and u'_C for various sets of joint loads P'_B and P'_C.

This behavior model can be used in an iterative procedure to design a structure. The properties of the structural components and configuration can be chosen and the response tested via the behavior model. If the response does not meet the chosen criteria, then the properties of the structural components and configuration must be changed and the analysis repeated.

Numerical Example

As an illustration, consider the specific design shown in Figure 3.5. According to Equation 3.21,

$$\begin{bmatrix} u'_B \\ u'_C \end{bmatrix} = \begin{bmatrix} 12 & -4 \\ -4 & 5.5 \end{bmatrix}^{-1} \begin{bmatrix} 5 \\ 8 \end{bmatrix}$$

Therefore,

$$\begin{bmatrix} u'_B \\ u'_C \end{bmatrix} = \begin{bmatrix} 0.11 & 0.08 \\ 0.08 & 0.24 \end{bmatrix} \begin{bmatrix} 5 \\ 8 \end{bmatrix} = \begin{bmatrix} 1.19 \text{ in.} \\ 2.32 \text{ in.} \end{bmatrix}$$

and $P'_A = -8u'_B - 1.5u'_C = -13$ lbs.

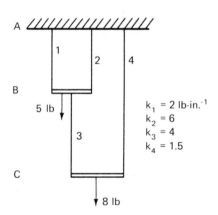

Figure 3.5

Compound Bar Example

The complete system solution is shown in Figure 3.6. If the loads are changed so that $P'_B = 6$ lbs and $P'_C = 10$ lbs, the new solution can be

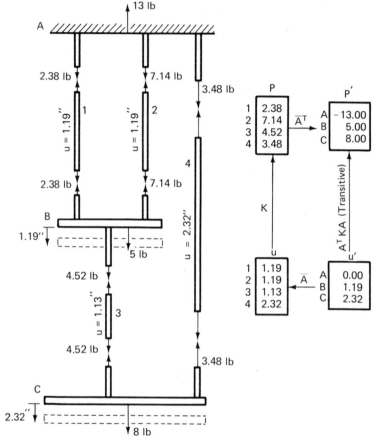

Figure 3.6

Solution for Compound Bar Example

computed as follows:

$$\begin{bmatrix} u'_B \\ u'_C \end{bmatrix} = \begin{bmatrix} 0.11 & 0.08 \\ 0.08 & 0.24 \end{bmatrix} \begin{bmatrix} 6 \\ 10 \end{bmatrix} = \begin{bmatrix} 1.46 \text{ in.} \\ 2.88 \text{ in.} \end{bmatrix}$$

and $P'_A = -8u'_B - 1.5u'_C = 16.00$ lbs.

Linear Graph Modeling of the Problem

A linear graph and topological matrix formulation of the compound bar problem is possible and is very useful as the logical basis for the automatic formulation of problems for solution by digital computer.

Referring to Figure 3.7, a directed linear graph abstraction of the schematic model is possible and leads to the derivation of the augmented incidence matrix $\bar{\mathbf{A}}$:

$$\bar{\mathbf{A}} = \begin{array}{c} \\ 1 \\ 2 \\ 3 \\ 4 \end{array} \begin{array}{ccc} A & B & C \\ \left[\begin{array}{ccc} -1 & 1 & 0 \\ -1 & 1 & 0 \\ 0 & -1 & 1 \\ -1 & 0 & 1 \end{array}\right] \end{array}$$

which is identical to the coefficient matrix $\bar{\mathbf{A}}$ in Equation 3.6. Thus, once the augmented matrix has been derived, the system model can be obtained according to Equation 3.12; i.e.,

$$\mathbf{P}' = (\bar{\mathbf{A}}^T \mathbf{K} \bar{\mathbf{A}}) \mathbf{u}'$$

Furthermore, since joint A is fixed, node A in the linear graph model may be designated as the datum node. The corresponding branch-node incidence matrix $\mathbf{A}$ will be as follows:

$$\mathbf{A} = \begin{bmatrix} 1 & 0 \\ 1 & 0 \\ -1 & 1 \\ 0 & 1 \end{bmatrix}$$

It can be easily shown that the final system response model in Equation

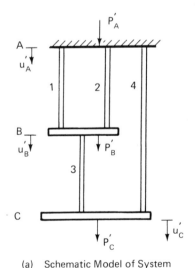

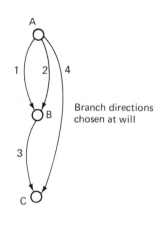

(a)   Schematic Model of System   (b)   Linear Graph of System

Figure 3.7

Linear Graph Model of Compound Bar Example

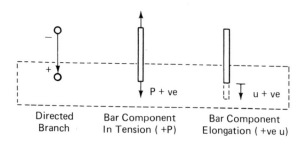

Figure 3.8

Linear Graph Sign Convention for Compound Bar Example

3.21 can be expressed as

$$\mathbf{u'} = (\mathbf{A}^T\mathbf{KA})^{-1}\mathbf{P'}$$ (3.22)

where

$$\mathbf{P'} = \begin{bmatrix} P'_B \\ P'_C \end{bmatrix} \text{ and } \mathbf{u'} = \begin{bmatrix} u'_B \\ u'_C \end{bmatrix}$$

Thus, given any compound bar system, it is possible to formulate a linear graph representation and derive the branch-node incidence matrix $\mathbf{A}$. The system response model can then be constructed according to Equation 3.22.

In order to make the graphical formulation general, it is necessary simply to devise a sign convention relating the linear graph branch directions to member forces and elongations, as in Figure 3.8, and to generate a consistent total system sign convention for nodal forces and displacements, as shown in Figure 3.7.

3.4 A LEVELING NETWORK MODEL

Consider a simple problem in topographic surveying. Figure 3.9 shows a network of five points well distributed within a dam site that is

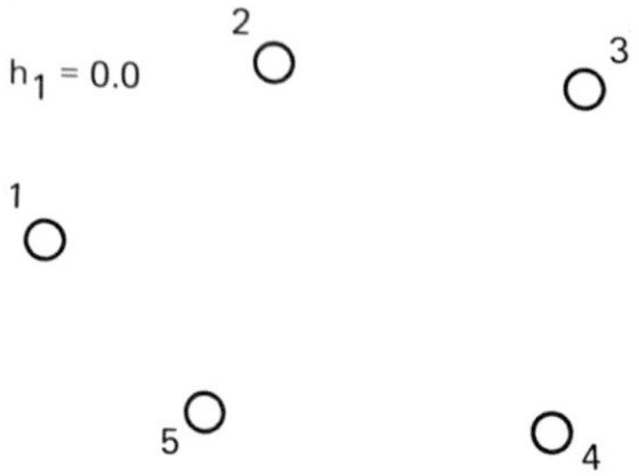

$h_1 = 0.0$

Figure 3.9

Network of Elevation Points

being surveyed and mapped. The elevation of point 1 has been determined by previous surveys. The problem is to determine the elevation of the remaining points relative to point 1; i.e., the elevation of point 1 will be assumed to be zero during the development of the network model.

Figure 3.10 illustrates one common leveling method used for mea-

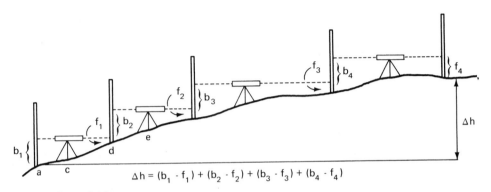

$$\Delta h = (b_1 - f_1) + (b_2 - f_2) + (b_3 - f_3) + (b_4 - f_4)$$

Figure 3.10

Leveling Method

suring the elevation difference between two points. Let b_i and f_i be a series of height measurements, as shown in the figure; then the elevation difference, Δh, can be computed as follows:

$$\Delta h = \sum_{i=1}^{n} (b_i - f_i) \tag{3.23}$$

The elevation h_2 of point 2 is then computed from the elevation h_1 of point 1 by the following formula:

$$h_2 = h_1 + \Delta h \tag{3.24}$$

The accuracy of the elevation difference, Δh, depends on the accuracy of the height measurements, b_i and f_i, and on the accuracy of the instruments used. Some errors invariably appear in engineering measurements; therefore, h_2 cannot be considered as the true elevation of point 2. In practice, the instruments and the leveling procedure are designed such that the only errors present in the measurements are small and unavoidable and result from limitations of the instrument and of the human observer. Moreover, in order to insure accurate determination of the elevation of the unknown points, the leveling network is designed so that the elevation of an unknown point can be determined from at least two different sets of elevation differences.

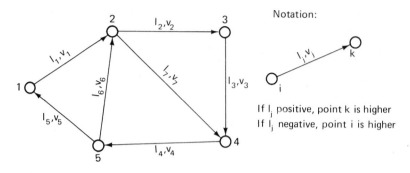

Figure 3.11

Linear Graph Model of a Leveling Network

For the purpose of this example, let the linear graph model in Figure 3.11 represent the leveling network used to determine the elevations of the unknown points in the dam site area. The branches represent the elevation differences that have been measured, the symbol l_j denotes the measured elevation difference between points i and k, and v_j is the most probable error in the measurement. The arrow is used to indicate the positive direction of l_j. If l_j is positive, the point at the head of the arrow is higher than the point at the tail of the arrow. If l_j is negative, then point k is lower than point i. It is necessary to find the most probable elevation at the unknown points from this set of measurements; the symbol h_j is used to denote the most probable elevation of point j. A mathematical model can be constructed to represent this leveling network as follows:

Identifying System Components

The system components can first be identified as follows:

Measured parameters: l_j for $j = 1$ to 7

Unknown parameters: h_i for $i = 2$ to 5

                            v_j for $j = 1$ to 7

System Constraint No. 1: System Compatibility

According to the notation in Figure 3.11, the following geometric relationships must obviously exist among the system components:

$$l_j - v_j = h_k - h_i \tag{3.25}$$

Since v_j is the error in the measurement l_j, it must be subtracted from the latter to yield the corrected elevation difference. One such equation is generated for each measured elevation difference (i.e., each branch on the linear graph in Figure 3.11), and the following set of equations results:

$$
\begin{aligned}
l_1 - v_1 &= h_2 \\
l_2 - v_2 &= -h_2 + h_3 \\
l_3 - v_3 &= -h_3 + h_4 \\
l_4 - v_4 &= -h_4 + h_5 \\
l_5 - v_5 &= -h_5 \\
l_6 - v_6 &= h_2 - h_5 \\
l_7 - v_7 &= -h_2 + h_4
\end{aligned}
\tag{3.26}
$$

By grouping all the unknown parameters at the left-hand side, the following set of equations results:

$$
\begin{aligned}
v_1 + h_2 & & &= l_1 \\
v_2 - h_2 + h_3 & & &= l_2 \\
v_3 \quad - h_3 + h_4 & & &= l_3 \\
v_4 \quad - h_4 + h_5 &= l_4 \\
v_5 \quad - h_5 &= l_5 \\
v_6 + h_2 \quad - h_5 &= l_6 \\
v_7 - h_2 \quad + h_4 &= l_7
\end{aligned}
\tag{3.27}
$$

In matrix notation, this may be expressed as follows:

$$
\begin{bmatrix} v_1 \\ v_2 \\ v_3 \\ v_4 \\ v_5 \\ v_6 \\ v_7 \end{bmatrix}
+
\begin{bmatrix}
1 & 0 & 0 & 0 \\
-1 & 1 & 0 & 0 \\
0 & -1 & 1 & 0 \\
0 & 0 & -1 & 1 \\
0 & 0 & 0 & -1 \\
1 & 0 & 0 & -1 \\
-1 & 0 & 1 & 0
\end{bmatrix}
\begin{bmatrix} h_2 \\ h_3 \\ h_4 \\ h_5 \end{bmatrix}
=
\begin{bmatrix} l_1 \\ l_2 \\ l_3 \\ l_4 \\ l_5 \\ l_6 \\ l_7 \end{bmatrix}
\tag{3.28}
$$

i.e.,
$$
\mathbf{V}_{(7\times1)} + \mathbf{A}_{(7\times4)}\mathbf{H}_{(4\times1)} = \mathbf{L}_{(7\times1)}
\tag{3.29}
$$

It can easily be seen from the linear graph in Figure 3.11 that the matrix $\mathbf{A}$ is, in fact, the branch-node incidence matrix. The matrices $\mathbf{V}$, $\mathbf{L}$, and $\mathbf{H}$ are column vectors of branch variables v_i, observed values l_i, and node variables h_i, respectively.

System Constraint No. 2: Behavior of the Measurement Errors

Equation 3.29 do not constitute a complete mathematical model of the problem, because there is not yet sufficient equations to solve for the unknowns. To complete the model, the characteristics of the measurement errors, v_i, must be defined. Assuming that only small unavoidable human errors are present in the measurements, and that the elevation differences, l_i, are all measured with *equal accuracy*, the following postulates may be used in this problem.

Choose any unknown point, say point 2. Suppose that the most probable elevations, h_3, h_4, and h_5, at points 3, 4, and 5, respectively, have been determined; that h_1 is assumed to be zero; and that the elevation differences from these points to point 2 have been measured to be l_2, l_7, l_6, and l_1. Thus, four different values of the elevation at point 2 can be computed as follows:

$$(h_2')_A = h_3 - l_2$$
$$(h_2')_B = h_4 - l_7$$
$$(h_2')_C = h_5 + l_6$$
$$(h_2')_D = l_1$$

Without knowing the actual error in the measurements, and since all measurements are equally accurate, the logical choice then is to compute h_2 as the mean of the four independently computed values:

$$h_2 = \tfrac{1}{4}[(h_2')_A + (h_2')_B + (h_2')_C + (h_2')_D]$$

i.e.,

$$4h_2 = h_3 + h_4 + h_5 + l_1 - l_2 + l_6 - l_7$$

Rearranging terms yields:

$$4h_2 - h_3 - h_4 - h_5 = l_1 - l_2 + l_6 - l_7$$

Applying the same logic at each of the remaining points yields the following:

Point 3 $-h_2 + 2h_3 - h_4 \qquad = l_2 - l_3$

Point 4 $-h_2 - h_3 + 3h_4 - h_5 = l_3 - l_4 + l_7$

Point 5 $-h_2 \qquad -h_4 + 3h_5 = l_4 - l_5 - l_6$

Putting all four of these equations into a single matrix equation yields

$$\begin{bmatrix} 4 & -1 & -1 & -1 \\ -1 & 2 & -1 & 0 \\ -1 & -1 & 3 & -1 \\ -1 & 0 & -1 & 3 \end{bmatrix} \begin{bmatrix} h_2 \\ h_3 \\ h_4 \\ h_5 \end{bmatrix} = \begin{bmatrix} l_1 - l_2 + l_6 - l_7 \\ l_2 - l_3 \\ l_3 - l_4 + l_7 \\ l_4 - l_5 - l_6 \end{bmatrix} \qquad (3.30)$$

i.e.,

$$\mathbf{B}_{(4\times4)}\mathbf{H}_{(4\times1)} = \mathbf{D}_{(4\times1)} \qquad (3.31)$$

It can be easily shown that the following relationships exist between the $\mathbf{B}$ and $\mathbf{D}$ matrices in Equation 3.31 and the $\mathbf{A}$ and $\mathbf{L}$ matrices in Equation 3.29:

$$\mathbf{B}_{(4\times4)} = \mathbf{A}^T_{(4\times7)}\mathbf{A}_{(7\times4)}$$

and

$$\mathbf{D}_{(4\times1)} = \mathbf{A}^T_{(4\times7)}\mathbf{L}_{(7\times1)}$$

Therefore, Equation 3.31 can be rewritten as

$$(\mathbf{A}^T\mathbf{A})\mathbf{H} = \mathbf{A}^T\mathbf{L} \qquad (3.32)$$

This set of constraint equations bears a striking similarity to Equation 3.29. In fact, by premultiplying both sides of Equation 3.29 by $\mathbf{A}^T$, the following expression results:

$$\mathbf{A}^T\mathbf{V} + (\mathbf{A}^T\mathbf{A})\mathbf{H} = \mathbf{A}^T\mathbf{L}$$

Thus, the above assumption about the behavior of the measurement errors implies that

$$\mathbf{A}^T\mathbf{V} = 0 \qquad (3.33)$$

i.e., the algebraic sum of the errors in the measurements to any node is equal to zero.

Complete System Model

A complete system model is obtained by grouping Equations 3.29 and 3.32:

Compatibility $\qquad \mathbf{V} + \mathbf{AH} = \mathbf{L} \qquad (3.29)$

Error Behavior $\qquad (\mathbf{A}^T\mathbf{A})\mathbf{H} = \mathbf{A}^T\mathbf{L} \qquad (3.32)$

Since Equation 3.32 involves four equations with four unknowns, the most probable elevations of the unknown points can be computed as follows:

$$\mathbf{H} = (\mathbf{A}^T\mathbf{A})^{-1}\mathbf{A}^T\mathbf{L} \qquad (3.34)$$

After the $\mathbf{H}$ matrix has been determined, it can be substituted in Equation 3.29 to yield the most probable measurement errors in matrix $\mathbf{V}$.

The $\mathbf{H}$ matrix presents the elevation of all points relative to the elevation of the datum at node 1. If the elevation of node 1 is not zero, then the elevation of node 1 must be added to each term in the $\mathbf{H}$ matrix to obtain the elevation of the other points.

An identical solution can be derived by formulating the problem with the least squares method, which is applied widely in the statistical analysis of engineering measurements.

In fact, the system model represented by Equations 3.29, 3.32, and 3.34 is completely general and is directly applicable to any leveling network that has one datum point. For any given leveling network, a linear graph model can be formulated and used to derive the coefficient matrix **A**. The system input will be the vector of measured elevation differences **L**, and the system output will be the vector of unknown elevations **H**. A direct solution can be obtained using the system model in Equation 3.34.

Numerical Example

Consider the leveling network represented by the linear graph in Figure 3.12.

$$
\mathbf{A} = \begin{array}{c} \\ 1 \\ 2 \\ 3 \\ 4 \\ 5 \\ 6 \end{array}
\begin{array}{ccc} 2 & 3 & 4 \end{array}
\begin{bmatrix} 1 & 0 & 0 \\ -1 & 0 & 1 \\ 0 & 1 & -1 \\ 0 & -1 & 0 \\ 0 & 0 & 1 \\ 1 & -1 & 0 \end{bmatrix} ; \quad
\mathbf{L} = \begin{array}{c} 1 \\ 2 \\ 3 \\ 4 \\ 5 \\ 6 \end{array}
\begin{bmatrix} 99.70 \\ 50.15 \\ 24.78 \\ -175.40 \\ 149.85 \\ -75.50 \end{bmatrix}
$$

$$
\mathbf{A}^T\mathbf{A} = \begin{bmatrix} 1 & -1 & 0 & 0 & 0 & 1 \\ 0 & 0 & 1 & -1 & 0 & -1 \\ 0 & 1 & -1 & 0 & 1 & 0 \end{bmatrix}
\begin{bmatrix} 1 & 0 & 0 \\ -1 & 0 & 1 \\ 0 & 1 & -1 \\ 0 & -1 & 0 \\ 0 & 0 & 1 \\ 1 & -1 & 0 \end{bmatrix}
$$

$$
= \begin{bmatrix} 3 & -1 & -1 \\ -1 & 3 & -1 \\ -1 & -1 & 3 \end{bmatrix}
$$

$$
\mathbf{A}^T\mathbf{L} = \begin{bmatrix} 1 & -1 & 0 & 0 & 0 & 1 \\ 0 & 0 & 1 & -1 & 0 & -1 \\ 0 & 1 & -1 & 0 & 1 & 0 \end{bmatrix}
\begin{bmatrix} 99.70 \\ 50.15 \\ 24.78 \\ -175.40 \\ 149.85 \\ -75.50 \end{bmatrix}
$$

$$
= \begin{bmatrix} -25.95 \\ 275.68 \\ 175.22 \end{bmatrix}
$$

$$(\mathbf{A}^T\mathbf{A})^{-1} = \begin{bmatrix} 0.5 & 0.25 & 0.25 \\ 0.25 & 0.5 & 0.25 \\ 0.25 & 0.25 & 0.5 \end{bmatrix}$$

$$\therefore \mathbf{H} = \begin{bmatrix} 0.5 & 0.25 & 0.25 \\ 0.25 & 0.5 & 0.25 \\ 0.25 & 0.25 & 0.5 \end{bmatrix} \begin{bmatrix} -25.95 \\ 275.68 \\ 175.22 \end{bmatrix} = \begin{bmatrix} 99.75 \\ 175.16 \\ 150.04 \end{bmatrix}$$

If the elevation of the datum node is 100.00 ft instead of 0.00 ft, a constant of 100.00 can now be added to each of the elevations computed above.

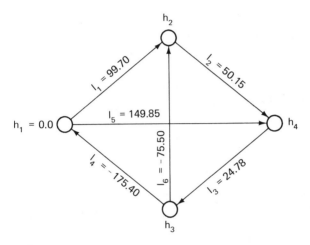

Figure 3.12

Leveling Network Example

3.5 A ROAD NETWORK IMPROVEMENT MODEL

To further illustrate the use of mathematical modeling, consider the road improvement problem which was analyzed in Section 2.6 using linear graphs. The problem statement and description are given in Section 2.6. The engineers in the highway department might proceed as follows:

1. Develop a mathematical model that will allow them to determine the maximum traffic flow of the existing network;

2. Develop a mathematical model that will allow them to determine which improvements to make in the network to allow a traffic flow of 6,000 vph and the cost of these improvements;

3. If the appropriation is less than the sum requested, develop a mathematical model that will allow them to determine which improvements to make such that the maximum traffic flow can be achieved for the amount of the appropriation.

Each of these models is developed below by first identifying the decision variables and then developing mathematical expressions relating these variables according to the system constraints.

Network Capacity Model

The network capacity model is developed for current flow capacity based on the following assumptions:

1. Traffic flow from A to B is independent and equal to flow from B to A;

2. Traffic flow direction on a specific road is known, as shown in Figure 2.16;

3. The traffic flow originating or terminating at the towns between A and B is insignificant.

Let
  ϕ_i = the actual flow in road i, i.e., branch flow;
  c_i' = the capacity of road i, i.e., branch capacity;
  ϕ_i' = imposed flow at node i, i.e., flow originating or terminating at node i; $\phi_i' = 0$ for $i = 2, 3, \ldots , 7$ from assumption 3 above.

Therefore, the maximum flow will occur in the network when ϕ_1' is a maximum. The value of ϕ_1' is limited, however, by two constraints:

1. The capacity of each road cannot be exceeded;
2. The flow at each node must be in equilibrium.

The capacity constraints on the existing roads can be determined from Figure 2.15. They are:

$$\text{for existing roads} \begin{cases} \text{four-lane} \begin{cases} 0 \leq \phi_1 \leq 2{,}000 \\ 0 \leq \phi_2 \leq 2{,}000 \\ 0 \leq \phi_3 \leq 2{,}000 \\ 0 \leq \phi_4 \leq 2{,}000 \end{cases} \\ \text{two-lane} \begin{cases} 0 \leq \phi_5 \leq 500 \\ 0 \leq \phi_6 \leq 500 \\ 0 \leq \phi_7 \leq 500 \\ 0 \leq \phi_8 \leq 500 \\ 0 \leq \phi_9 \leq 500 \\ 0 \leq \phi_{10} \leq 500 \end{cases} \end{cases} \quad (3.35)$$

In matrix form,

$$\mathbf{0}_{(10\times1)} \leq \boldsymbol{\Phi}_{(10\times1)} \leq \mathbf{C}_{(10\times1)} \quad (3.36)$$

where

$\boldsymbol{\Phi}$ = a column matrix of road traffic flow;

$\mathbf{C}$ = a column matrix of road flow capacities and represents a physical behavior property of the road components.

The second constraint means that the actual road flows, ϕ_i, at the nodes must balance at road junctions because of continuity requirements.

If the flow into a node is considered as positive, then at source A, node 1, the continuity equation must relate the network nodal flow, ϕ_1', to the road flows ϕ_1, ϕ_5, and ϕ_6. Furthermore, the direction of traffic flows is known, hence:

$$\phi_1' - \phi_1 - \phi_5 - \phi_6 = 0 \quad (3.37)$$

or

$$\phi_1 + \phi_5 + \phi_6 = \phi_1' \quad (3.38)$$

Similarly, for

$$\begin{array}{lll}
\text{Node 2} & -\phi_2 + \phi_5 & = 0 \\
3 & +\phi_1 - \phi_7 & = 0 \\
4 & -\phi_3 + \phi_6 - \phi_8 = 0 \\
5 & +\phi_2 - \phi_9 & = 0 \\
6 & +\phi_3 - \phi_4 + \phi_7 = 0 \\
7 & +\phi_8 - \phi_{10} & = 0 \\
8 & +\phi_4 + \phi_9 + \phi_{10} = \phi_8'
\end{array} \quad (3.39)$$

Equation 3.39 can be expressed in matrix form as

$$\mathbf{A}^T_{(7\times10)}\boldsymbol{\Phi}_{(10\times1)} = \boldsymbol{\Phi}'_{(7\times1)} \quad (3.40)$$

where

$\mathbf{\Phi}'$ = the column matrix of imposed nodal flows exclusive of ϕ_1'
$\mathbf{A}$ = the branch-node incidence matrix for Figure 2.16.

The flow capacity model can be summarized in matrix form as:

Maximize

$$\phi_1' = \phi_1 + \phi_5 + \phi_6 \qquad (3.41)$$

subject to

$$\mathbf{0}_{(10\times1)} \leq \mathbf{\Phi}_{(10\times1)} \leq \mathbf{C}_{(10\times1)} \qquad (3.36)$$

$$\mathbf{A}^T_{(7\times10)}\mathbf{\Phi}_{(10\times1)} = \mathbf{\Phi}'_{(7\times1)} \qquad (3.40)$$

Equations 3.41, 3.36, and 3.40 constitute a linear programming model with ten flow variables $\phi_1, \ldots, \phi_{10}$. The values of these variables must be determined to obtain the solution. Computer programs are available on most computer systems that can solve these equations in seconds of computer processor time.

Improvement Cost Model

If the solution for the network capacity model indicates that the total flow capacity of the network is less than the demand capacity of 6,000 vph, then the next step is to construct a model that will indicate which improvements should be made to the network and the cost of these improvements to satisfy the demand.

During this step, the objective is to find the minimum cost of increasing the network capacity to 6,000 vph. In order to incorporate the network flow capacity as a function of new lane and road construction, it is necessary to introduce decision variables for the new construction effects on the capacity constraints.

Let $d_{i,1}$ be a decision variable assigned to an existing road i and let it take a value of either 0 or 1 with the following meaning:

$$d_{i,1} = 0 \quad \text{do not add any new lanes;}$$

$$d_{i,1} = 1 \quad \text{add two lanes to road } i.$$

Furthermore, let $d_{j,2}$ be the decision variable assigned to a proposed road j. Similarly, $d_{j,2}$ can take on either 0 or 1 with the following meaning:

$$d_{j,2} = 0 \quad \text{do not build proposed road } j;$$

$$d_{j,2} = 1 \quad \text{build a new two-lane road } j.$$

Thus, the additional decision variables for this problem can be symbolically represented by $d_{5,1}$, $d_{6,1}$, $d_{7,1}$, $d_{8,1}$, $d_{9,1}$, $d_{10,1}$, $d_{11,1}$, and $d_{11,2}$. For the proposed road 11, the decision variables $d_{11,1}$ and $d_{11,2}$ have the following combined meaning:

$$d_{11,1} = 0 \text{ and } d_{11,2} = 0 \quad \text{no construction;}$$
$$d_{11,1} = 0 \text{ and } d_{11,2} = 1 \quad \text{build two lanes;}$$
$$d_{11,1} = 1 \text{ and } d_{11,2} = 1 \quad \text{build four lanes.}$$

According to the above definition, $d_{11,1}$ cannot be equal to 1 when $d_{11,2} = 0$.

The objective, then, is to minimize the sum of the costs for new road construction. The new construction costs for existing road i become \$(mileage road i)$Rd_{i,1}$ and the costs for road 11 become \$(mileage road 11) $(1.5Rd_{11,2} + 0.8Rd_{11,1})$. Thus the objective is

Minimize

$$(20d_{5,1} + 20d_{6,1} + 20d_{7,1} + 30d_{8,1} + 30d_{9,1} + 20d_{10,1} + 60d_{11,2}$$
$$+ 32d_{11,1})R \quad (3.42)$$

subject to capacity constraints:

$$
\text{for existing roads}
\begin{cases}
\text{four-lane}
\begin{cases}
0 \le \phi_1 \le 2{,}000 \\
0 \le \phi_2 \le 2{,}000 \\
0 \le \phi_3 \le 2{,}000 \\
0 \le \phi_4 \le 2{,}000
\end{cases} \\
\\
\text{two-lane}
\begin{cases}
0 \le \phi_5 \le 500 + 1{,}500d_{5,1} \\
d_{5,1} = 0 \text{ or } 1 \\
0 \le \phi_6 \le 500 + 1{,}500d_{6,1} \\
d_{6,1} = 0 \text{ or } 1 \\
0 \le \phi_7 \le 500 + 1{,}500d_{7,1} \quad (3.43) \\
d_{7,1} = 0 \text{ or } 1 \\
0 \le \phi_8 \le 500 + 1{,}500d_{8,1} \\
d_{8,1} = 0 \text{ or } 1 \\
0 \le \phi_9 \le 500 + 1{,}500d_{9,1} \\
d_{9,1} = 0 \text{ or } 1 \\
0 \le \phi_{10} \le 500 + 1{,}500d_{10,1} \\
d_{10,1} = 0 \text{ or } 1
\end{cases}
\end{cases}
$$

$$\text{for planned road 11} \quad \left\{ \begin{array}{l} 0 \leq \phi_{11} \leq 500d_{11,2} + 1{,}500d_{11,1} \\ d_{11,1} = 0 \text{ or } 1 \\ d_{11,2} = 0 \text{ or } 1 \\ d_{11,1} \leq d_{11,2} \end{array} \right. \qquad (3.44)$$

and subject to continuity conditions that are the same as in Equation 3.39 except for nodes 2 and 6, which now become

$$\begin{array}{ll} \text{Node 2} & -\phi_2 + \phi_5 - \phi_{11} = 0 \\ \quad\;\; 6 & +\phi_3 - \phi_4 + \phi_7 + \phi_{11} = 0 \end{array} \qquad (3.45)$$

and subject to the constraint that the total network flow is equal to 6,000; i.e.,

$$\phi_1 + \phi_5 + \phi_6 = 6{,}000 \qquad (3.46)$$

This results in an integer linear programing model with eleven flow variables, $\phi_1, \ldots, \phi_{11}$ and eight integer decision variables.

The solution of this model for the flow variables and the decision variables will provide the cost of increasing the network capacity to 6,000 vph and will indicate which roads should be improved. This information will then provide the basis on which to make a request for the money to construct the needed roads.

Budget Constraint Model

If the total amount of money requested for new construction by the highway department is not appropriated, then, the department must work within the amount appropriated. Therefore, the objective now becomes maximizing the flow capacity of the network by increasing the capacity of some of the roads subject to the budget constraint. Hence, the objective function is

Maximize

$$\phi_1' = \phi_1 + \phi_5 + \phi_6 \qquad (3.47)$$

subject to the constraints of Equations 3.43, 3.44, and 3.39, as modified by Equation 3.45. It is also subject to the constraint that

$$(20d_{5,1} + 20d_{6,1} + 20d_{7,1} + 30d_{8,1} + 30d_{9,1} + 20d_{10,1} + 60d_{11,2}$$
$$+ 32d_{11,1})R \leq S \qquad (3.48)$$

where S is the amount appropriated for the network improvement. This model is very similar to the network capacity model. The only difference is that the capacity and continuity constraints now include the provision for the proposed roads, and there is an additional constraint, Equation 3.48, due to the budget limitation.

Determining Optimum Direction of Flows

Mathematical modeling may also be used to determine the optimum direction of flow through the branches in a network. Suppose that branch i in a flow network has a capacity of c_i, as shown in Figure 3.13a.

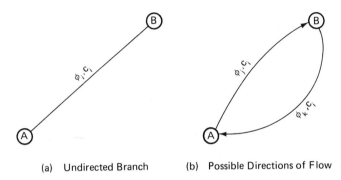

(a)   Undirected Branch   (b)   Possible Directions of Flow

Figure 3.13

Linear Graph Model When Flow Direction Is Unknown

Furthermore, suppose that the flow can be either from node A to node B or from B to A; but, simultaneous flow in both directions is not possible. Part of the system, then, is to determine the direction of flow that results in maximum flow

Since flow in either direction is possible, the undirected branch in Figure 3.13a can be replaced by two branches directed in opposite directions, as shown in Figure 3.13b. Let d_j be a decision variable that has a value either of 1 or 0; 1 means that there is a flow along directed branch j, and 0 means that there is no flow through the branch. The following equations can thus be written to describe the capacity constraints:

$$0 \leq d_j \phi_j \leq c_i$$
$$d_j = 1 \text{ or } 0$$
$$0 \leq d_k \phi_k \leq c_i \tag{3.49}$$
$$d_k = 1 \text{ or } 0$$

Since simultaneous flow in both directions is not possible, d_j and d_k cannot both be equal to 1. This condition can be expressed as follows:

$$d_j + d_k \leq 1 \tag{3.50}$$

Thus, Equations 3.49 and 3.50 together describe the flow condition in the original branch i. Note, however, that the flow equations in

Equation 3.49 are nonlinear equations since the terms $d_j\phi_j$ and $d_k\phi_k$ are products of two unknown variables. The resultant mathematical model for the network will be a nonlinear programing model.

3.6 MATHEMATICAL MODELING OF SYSTEMS

All system models must portray in some manner the number and behavior of its components, the system structure of the component interactions, and system conditions introduced by the exposure of the system to physical laws and requirements imposed by organizational considerations. In addition, the relationship and use of the model by the modeler in the design process may require that the model exhibit decision aspects.

The essential difference between the mathematical and graphical modeling of systems is that the mathematical model must provide explicit structural statements in its formulation, whereas the graphical model analysis can build on an existing graph structure. Consequently, the mathematical model and the individual equations it uses often exhibit the mathematical characteristics of the system graph structure. The use of mathematical models is usually more convenient than a graphical analysis when analyzing large systems.

All the examples developed in this chapter explicitly utilize and exhibit graphical structural properties in some segments of the models. Thus, if the connectivity of branches and nodes is used to define the structural interaction of the components, then the branch-node incidence matrix $\mathbf{A}$ or its transpose $\mathbf{A}^T$ will appear in the mathematical formulation.

All equations expressing compatibility between branch and node quantities will exhibit the following form:

$$(\text{branch vector}) = \mathbf{A} \ (\text{node vector})$$

For example, in the compound bar problem, the relationship between branch elongations and joint displacements is expressed as:

$$\mathbf{u} = \mathbf{A}\mathbf{u}'$$

In the leveling network problem, the relationship between elevation differences and elevation is expressed as $(\mathbf{V} + \mathbf{L}) = \mathbf{A}\mathbf{H}$.

Similarly, all equations involving equilibrium or continuity in which branch quantities are summed up at a node, will exhibit the form:

$$(\text{node vector}) = \mathbf{A}^T \ (\text{branch vector})$$

For example, in the compound bar problem, equilibrium of branch and nodal forces at the joints is expressed as

$$\mathbf{P}' = \mathbf{A}^T\mathbf{P}$$

For the survey problem, the distribution of errors in the measurements yields

$$A^T V = 0$$

For the highway problem, the continuity of flow along highways and through intersections gives

$$A^T \Phi = \Phi'$$

Although the examples exhibit this modeling similarity, they represent the solution of engineering problems that differ in complexity. In the case of the compound bars, the components and structures are well defined. The constraints must conform to the laws of science that apply to the problem. However, the purpose of the system is not really defined in terms of what objectives or goals the system must fulfill.

The second example, involving a surveying network, focuses on an observational and measurement problem. The formulation requires an explicit recognition of the fact that measurements are always inexact and an implied error behavior is built into the model.

The highway problem, presented in the third example, represents a system with an expressed purpose that must be fulfilled. It also illustrates that the exact form of the equations in the model will depend on how the problem is stated. If the objective is to determine the maximum network capacity, one model will be appropriate, but if the objective is to minimize cost of network improvement, a model with slightly different equations becomes appropriate. In reality, the problem could be expanded to involve other considerations, such as construction schedule and roadway maintenance. Thus, in this case, all the problem components have not been fully defined at this point. The consideration of these other aspects involves the management of the system and are discussed in Chapter 10.

Note that regardless of the level of complexity of the problem, mathematical modeling can be utilized. In planning and designing engineering systems, mathematical modeling can serve as a valuable analytical tool.

3.7 PROBLEMS

P3.1. Develop descriptive, symbolic, schematic, and mathematical models for each of the following situations. To what extend would your model formulation change if your purpose is to:

1. describe the physical situation,
2. design and manufacture the facility,

3. use the facility.
   a. A man standing on a ladder that makes an angle of 20°
      with a vertical wall;
   b. The flow of electricity through an insulated cable;
   c. A 300 lb weight resting on the free end of a horizontal
      cantilever;
   d. A steel cable supported at both ends;
   e. A vehicle moving along a curve of 900 ft radius at a speed
      of 60 mph;
   f. The foyer entrance to a high-rise building.

P3.2. Identify a simple design problem associated with each of
the following systems. In each case, specify the design and response

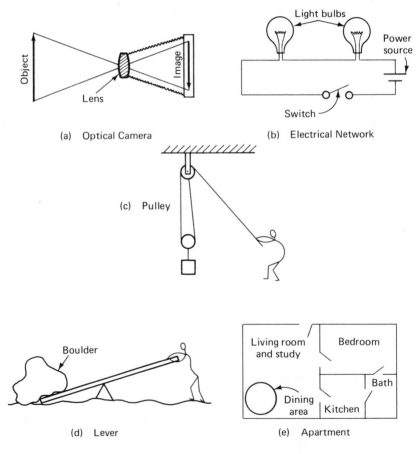

Figure **P3.2**

variables and develop mathematical expressions to relate these variables to the relevant system parameters.

P3.3. Develop a network capacity model, an improvement cost model, and a budget constraint model for the road network described in Section 2.6 if the following additional features must be considered:

1. Because of surface deterioration the capacities of roads 4, 7, and 8 are reduced to 1,700, 300, and 300 vehicles per hour, respectively. The capacity of road 4 can be increased at the rate of 300 vehicles per hour for each $0.5R$ per mile spent for road repair. The capacities of roads 7 and 8 can each be increased at the rate of 200 vehicles per hour for each $0.1R$ per mile spent for road repair. The improved capacity of each road cannot be greater than its original capacity.

2. The total budget for road repair and new road construction is S.

P3.4. A plant production line consists of five processing units, as shown in Figure P3.4a. Processing unit P4 is an assembly process requiring the assembly of two items from P3 on line B and one item from P2 in line A. All other processing units perform operations on individual items.

The production from each processing unit measured in terms of items processed depends on both the input supply to the unit and on the age of the processing unit in terms of actual (or effective) months of production spent in the production line. In addition, while the productive output of a unit decreases with time, the percentage of defective items produced increases with time, so that the effective production of acceptible items is a function of both unit age and quality.

Assume that the maximum output capacities of each unit decrease linearly with time and that the percentage of defective items increases parabolically with time, as illustrated in Figure P3.4b and quantified in Table P3.4.

Thus in any specific unit the number of defective items produced is given by $(c + dx^2)$, where x is the actual or effective age (in months) of the process unit in the production line.

The supply rates I_1 and I_2 for lines A and B have been set at 120 and 140 (thousands per month), although the factory could deliver at the maximum rates of 150 (thousands per month) to both lines.

1. What is the monthly output of acceptable finished products for the first 6 months of operation?

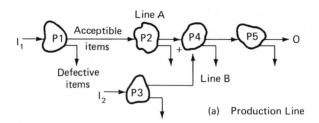

(a)   Production Line

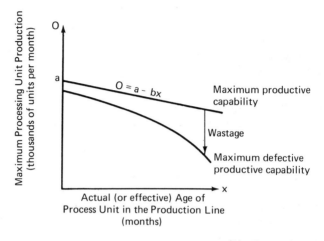

(b)   Output Capacity for
Each Production Unit

Figure P3.4

Table P3.4

	a	b	c	d	INITIAL AGE OF MACHINE (MONTHS)
P1	120	7	0.85	0.5	1
P2	100	6	0.75	0.5	0
P3	150	7	0.95	0.5	2
P4	80	3	0.75	0.5	1
P5	80	4	0.95	0.5	1

2. What are the optimal values of I_1 and I_2 for this production rate and what bottlenecks exist?

3. What single processing unit should be replaced now at the start of the production run? Find the new monthly output for the first 6 months and the optimal values of I_1 and I_2.

P3.5. Develop a mathematical formulation for the broad characteristics of a high-rise building with a rectangular cross section.

1. What would you choose for an objective and how would you incorporate into your model the following considerations?
   a. local zoning ordinances restricting the building height and percentage of ground cover,
   b. floor space required for tenants and services,
   c. structural, mechanical, and elevator costs that increase with building height,
   d. differential building surface and heating costs for the east-west and north-south faces.
2. How could you incorporate into the model considerations relating to
   a. rental income as a function of type of tenant,
   b. maintenance cost,
   c. total cost?

P3.6. The following linear graph represents a water pipe network. The lengths of the pipe sections are shown in the graph, and all pipes are 12″ in diameter. The flow (q_i) along the ith pipe may be related to the pipe diameter (d_i), length of pipe (L_i), and head loss along the pipe (h_i) by the following equation

$$q_i = 44.8 \left(\frac{d_i^{2.5}}{L_i^{0.5}} \right) h_i^{0.5}$$

1. What are the three different sets of physical constraints that must be satisfied by the above network in a steady state flow condition?
2. Develop a mathematical model of the system by deriving equations to describe all the physical constraints.
3. By means of the mathematical model developed in (2) above, compute the flows (q_i) and the hydrostatic head at joints B and C.

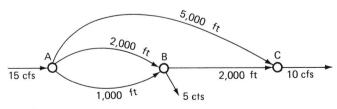

Figure P3.6

P3.7. Develop a mathematical formulation for the compound bar problem discussed in Section 3.3, which includes the following considerations:

1. Bar lengths, sizes, and material types are to be explicitly included.
2. Costs must be considered as a function of material type, quantity used, and joint costs as a function of bar perimeter attached to the joint.

How is the solution process affected, and what might be the objectives and constraints, that you must now consider?

P3.8. The models that have been developed above are the types that are useful for components in the need model shown in Figure 1.11. What is needed is a model that incorporates several need models into a general model for the entire system, as shown in Figure 1.12. How would you approach the problem of combining the types of models developed above into the general model for the entire system area of a problem?

4

OPTIMIZATION

The 17.6 mile Chesapeake Bay Bridge-Tunnel project was selected as the Out-standing Civil Engineering Achievement of 1965. The bridge-tunnel crosses over and under the open sea linking the state of Virginia to the north and south. The project combines $12\frac{1}{2}$ miles of low-level trestle, four man-made islands raised from the bottom of the bay, two tunnels that were sunk into place, a high and a medium-level bridge, and several miles of approach roads. The mile-long tunnels were built to meet navigation requirements of ocean going vessels. (Courtesy of the American Society of Civil Engineering)

4.1 THE OPTIMIZATION PROCESS

In most engineering problems, many feasible solutions exist. It is the professional responsibility of the engineer to seek out the best possible solution according to the goals and objectives for the total system. The processes by which the best solution is determined is called "optimization."

Optimization is an essential and necessary phase of the design and planning process. After the problem environment has been carefully defined and some initial solutions have been found, system design models are developed to describe the design characteristics and the interactions of the system requirements. These models may be graphical, mathematical, or physical. They provide a vehicle to analyze the behavior and response of the design when it is exposed to a wide range of environmental conditions. As a result of the understanding gained from these analyses, design modifications may be made or completely new designs may be needed. This cyclic process of design and analysis is reiterated until a set of most promising solutions to the system problems emerges. These solutions must then be evaluated and ranked according to how well they fulfill the goals and objectives that have been established for the proposed system. During the ranking and evaluation phase, new understanding may be gained and new system constraints may be discovered. It is then necessary to return again to the design and analysis cycle.

Thus, the optimization process is a cyclic process that consists of a continuous interplay of design, analysis, and ranking of alternative solutions. Figure 1.21 illustrates this basic principle. Within the framework of the systems approach, several methods and procedures have been developed to aid the decision-maker in his search for the optimum solution. The methods of linear graphs and mathematical modeling described in Chapters 2 and 3, respectively, help to describe the interactions of system components and to provide a vehicle for studying the input-output response characteristics of the systems. In later chapters, more specific system techniques will be discussed. The remainder of this chapter will be devoted to illustrating the fundamental concepts of optimization.

4.2 MOTIVATION AND FREEDOM OF CHOICE

For optimization to be possible, the engineer must have the desire or at least the incentive to find the best solution. This motivation comes from the need to efficiently utilize materials and resources to improve the quality of life. The degree of success in optimization is often governed by the extent to which the designer is motivated to seek an optimum solution.

In addition, optimization is possible only if a freedom of choice exists. The problem statement may be so restrictive that there is no opportunity for choice or selection; i.e., there may be only one feasible solution to the problem as stated. As the problem constraints are relaxed, the number of feasible solutions increases and the task of optimization becomes correspondingly more complex. It is important, therefore, that during the problem definition phase, the system constraints imposed must be truly representative of the goals and objectives of the problem. Unnecessary system constraints may disqualify an otherwise attainable optimum solution from further consideration.

For example, consider the design problems stated below:

Design Problem 4.1   Design a plane steel truss to span a distance of 120 ft. The truss must have the configuration and dimensions shown in Figure 4.1a. It supports a vertical load of 30 kips (1 kip = 1000 lb.) at joints 2 and 3. A fixed support is to be used at joint 4 and a roller support at joint 1. The steel should have a Young's modulus of elasticity of $E = 30,000$ kips/sq in. Members in tension must not be subjected to stresses greater than 20 kips per sq in. and members in compression must not be subjected to stresses greater than 10 kips per

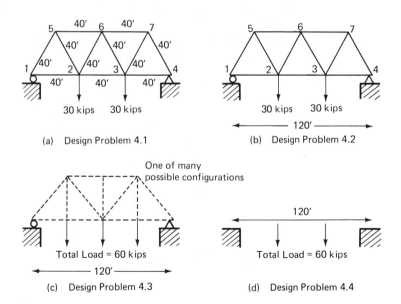

(a) Design Problem 4.1

(b) Design Problem 4.2

(c) Design Problem 4.3

(d) Design Problem 4.4

Figure 4.1

Influence of Constraints on Optimization Problems

sq in. The maximum joint deflection must be less than 1.3 in. and the total weight must not exceed 2.2 tons. Steel members are available in the following cross-sectional areas and costs: 2, 3, 4, 5, and 6 sq in. at a cost of \$210, \$220, \$240, \$260, \$280, and \$300 per member, respectively. The design must result in a minimum cost structure.

It is apparent that the primary task of the design engineer in this problem is to determine the lowest cost combination of cross-sectional areas for the members. The shape of the truss, the linear dimensions of the members, and the type of steel are all specified. Neither inventiveness nor originality is being required of the engineer, even though there may be several combinations of cross-sectional areas that satisfy the design constraints.

Design problems 4.2, 4.3, and 4.4 stated below show a progressive relaxation of the design constraints in the above truss problem.

Design Problem 4.2 Design a plane truss to span a distance of 120 ft. The truss must have the configuration shown in Figure 4.1b with a vertical load of 30 kips at joints 2 and 3. A fixed support is to be used at joint 4 and a roller support at joint 1. The steel should have a Young's modulus of elasticity of $E = 30,000$ kips/sq in. Members in tension must not be subjected to stresses greater than 20 kips per sq in. and members in compression must not be stressed greater than 10 kips per sq in. The maximum joint deflection must be less than 1.3 in. and the total weight must not exceed 2.2 tons. The cost of the structure is to be kept to a minimum. The following lengths and areas of members are available:

MEMBER LENGTH, FT	MEMBER AREA, SQ IN.	NUMBER OF MEMBERS
72.1	3, 4, 5, 6, 7	2 of each area
67.1	3, 4, 5, 6, 7	2 of each area
60.	3, 4, 5, 6, 7	2 of each area
50.	2, 3, 4, 5, 6	6 of each area
40.	2, 3, 4, 5, 6	12 of each area
30.	2, 3, 4, 5, 6	6 of each area

The costs of these members are as follows:

	COST IN DOLLARS FOR EACH MEMBER					
Length AREA	2 *sq in.*	3	4	5	6	7
72.1 ft	650	670	690	720	760	800
67.1	450	465	490	510	535	650
60	380	390	400	415	430	450
50	300	310	320	340	360	380
40	210	220	240	260	280	300
30	110	120	140	160	180	200

In this problem, the shape of the truss and type of steel are still specified, but the designer now has more freedom in choosing the dimensions of the members.

Design Problem 4.3   Design a plane steel truss to span a distance of 120 ft. The truss can be of any configuration and dimensions. A total vertical load of 60 kips is to be evenly distributed at the joints in the same level as the two end supports. The truss must meet all the other design constraints stated in design problem 4.2 and at minimum cost (see Figure 4.1c).

The shape of the truss and the exact location of the loadings are no longer specified. The engineer can now choose any truss configuration and member sizes as long as they satisfy the constraints on axial stress, weight, and vertical deflection.

Design Problem 4.4   Design a minimum cost structure to span a distance of 120 ft and support a total vertical load of 60 kips such that the maximum deflection in the plane of the supports is not greater than 1.3 in. (see Figure 4.1d).

This problem statement presents the design engineer with complete freedom to choose the type of structure, the type of materials, and the distribution of the vertical loads.

It should be obvious that the optimum design for design problem 4.1 is not necessarily the optimum solution for design problem 4.4, or vice versa. It is also apparent that the level of competance required of the designer increases with the degree of freedom in the problem.

4.3 GOALS, OBJECTIVES, AND CRITERIA

The purpose of the optimization process is to determine the system that allows the decision-maker to most nearly achieve his stated goals and objectives. The degree to which he achieves these goals and objectives must be measured by a specific set of criteria. Therefore, it is absolutely essential that the goals, objectives, and criteria of the problem be clearly defined during problem definition.

The goals must truly reflect the ultimate purposes of those who have valid direct or indirect interests in the problem. For example, consider again the problem of designing a high-rise building (HRB), which was discussed in Chapter 1. Among the many different interested parties, the most important ones are the owner, the tenants, and the general public, which is represented by the city administration. The primary goal of the owners may be to obtain a profit large enough to justify the venture; the tenants may be primarily interested in acquiring an attractive living or business environment for a reasonable cost; and the city administration may be primarily interested in improving the economic and social well-being of the city. These goals are not mutually independent. The tenants may or may not already have been identified during the design of the HRB, but it is apparent that if the goals of the owners are to be achieved, the interests of the prospective tenants must be taken into serious consideration. Similarly, the well-being of the city strongly affects the lives of tenants and, therefore, the future prospects of the HRB.

The problem objectives specify which characteristics of the system are to be optimized in order to achieve the goals. Objectives are commonly related to such factors as cost, profit, weight, speed, time, artistic features, color, shapes, forms, and efficiency. In order to achieve the HRB owner's goals of obtaining a large profit, some obvious objectives are:

1. To minimize the capital, maintenance, and operational cost, and

2. To maximize the total income from the HRB.

For the tenants' goals, the objectives may be:

1. To maximize the attractiveness of the living and/or business environment, both within and around the HRB,

2. To maximize the safety of the building, and

3. To maximize the convenience of entering and leaving the building.

For the city administration, the objectives may be to maximize the economic gain of the city while minimizing financial and operational burdens imposed on the city by the HRB. Again, these goals are not mutually exclusive, but conflict with, as well as complement, one another.

Finally, the criteria are the set of parameters used to measure how optimum a solution is with respect to the goals and objectives. For the owner's objectives of minimizing cost, the criteria may be expressed in terms of the capital investment in dollars, the interest rate for loans, inflation rate, city taxes, maintenance cost, and operational cost. For maximizing income, the criteria may include rental rates in dollars, occupancy rate, tenants' income levels, and expected life of the HRB. Possible criteria for other objectives in this HRB example are listed in Table 4.1.

Delineating the goals, objectives, and criteria, as shown in Table 4.1, is the necessary first step toward achieving an optimum design. For

Table 4.1

Possible Goals, Objectives, and Criteria for an HRB Project

GOALS	OBJECTIVES	CRITERIA
1) Maximize profit from investment;	1) Minimize cost	Capital investment in dollars; operational cost; maintenance cost; interest rate; city taxes
	2) Maximize income	Rental rates; occupancy rate; tenant's income levels; life of HRB
2) Maximize tenant satisfaction	3) Maximize safety	Structural safety factor against natural hazards such as storms, tornadoes, earthquakes, etc.; fire hazards and fire escapes; city building codes; design structural loading; emergency procedures
	4) Provide attractive living and/or business environment for tenants	Functional needs; scenic views; shopping and recreational facilities; personal security; transportation of goods and people; noise; proximity to employment; playgrounds and parks; churches and schools; prestige and pride; police protection
	5) HRB to be architecturally attractive	Conformity to surrounding buildings and environments; modern, traditional, or classic; degree of inventiveness; color; shape
3) Contribute to the well-being of the city	6) Maximize economic gain for the city	Creation of new industry and employment; improve urban housing environment
	7) Minimize financial and operational burden on the city	Increased demands for water and power supply, sewage disposal and processing facilities; impact on neighborhood traffic; demands for parks and recreational facilities; impacts on local schools

any given engineering problem, whether it be simple or complex, it may be extremely difficult to identify and state all the relevant goals, objectives, and criteria at the very beginning of the design process. However, as the problem is being studied and alternate designs are being developed, new understanding of the problem may result in new criteria, new objectives, or even new goals.

In addition, since the goals, objectives and criteria inevitably conflict, the design engineer must also carefully rank their relative importance for the purpose of trade-offs in the optimization process. The goals and objectives in Table 4.1 are listed in the order of importance from the viewpoint of the owners. Where conflicts arise, the design engineer must decide on the solution that best resolves the conflicting issues. In order to do this, the ranked value system may be utilized to establish trade-off equations among the conflicting elements. However, the establishment of trade-off equations requires actual quantification of the relative importance of the goals, objectives, and criteria.

In general, the success of optimization depends on the ability of the design engineer to:

1. Structure the hierarchial orders of goals, objectives, and criteria; and
2. Define a value system that establishes priority and preference.

On this basis, the engineer will undoubtedly be faced with a spectrum of problems in which there will be considerable variation in order and preference. The methods or techniques that are utilized for determining optimum solutions must recognize this aspect of the problem definition.

As a second example, consider the problem that deals with the design and location of a flood control dam on a river. A decision has been made to build a dam in a particular location; thus, the goal is set. It must be understood that the decision to build the dam in the first place is part of a larger scope and involvement.

Two objectives that immediately arise are:

1. Maximize the control of floodwaters of the river, and
2. Minimize the damage to the environment of the area.

The initial examination of these two objectives reveals that they are in direct conflict and both cannot be fulfilled.

The criteria for measuring the objectives also present a contrasting situation. In order to measure the flood control, a dollar value could be assigned to losses or potential losses from flooding. In essence, a measure of flood control is achieved. However, the damage to the environment is not so easily defined and, in some cases, it may be difficult to assess damage and consequences over a period of time. Nevertheless, they are important considerations in terms of the project.

Some type of trade-off must be established if a compromise is to be reached and the dam is to be ultimately built. In essence, increased cost of flood damage must be accepted in order to preserve the natural habitat of the area. Unless this trade-off can be achieved, the goal cannot realistically be attained and another solution to the larger problem must be sought. A choice must be made in such a problem, and the choice is an important part of the optimization problem.

4.4 OPTIMUM

An optimum solution is usually defined as the technically best solution that is achieved without compromising any goals and objectives. It represents the idealistic solution that allows all the goals and objectives to be attained. In reality, such a solution rarely exists. The nature of engineering problems invariably involves conflicting interests. Due to the state of the art in science and technology, many design factors and system constraints are not well understood; and many value criteria, such as social relevance, quality of life, ecology, and attractiveness, cannot be easily quantified or defined. The design and planning of engineering systems are always conducted within limited manpower, equipment, financial resources, and time. Therefore, the optimization process produces only an "optimal" solution, that is the best achievable within the design and technological constraints.

4.5 SUBOPTIMIZATION

Idealistically, all components of a system must be optimized with respect to the goals, objectives, and criteria of the total system. To do so will require a perfect understanding of the behavior of each component within the system structure, as well as its effects on the response characteristics of the total system. This requirement, in turn, calls for a single designer who has perfect understanding of every component within the system. A team of designers from various disciplines cannot perform the same function of a know-everything designer because of the difficulty in communicating technical information among different disciplines. Obviously, such requirements can never be met in complex engineering problems. Therefore, in practice, a design problem is often subdivided into component parts under the responsibility of several design teams. The quality and performance of these component parts are then controlled by sets of specifications that define the goals and objectives for each component. The process of optimizing a component of a system according to a subset of goals, objectives, and criteria is called suboptimization.

For example, in the HRB problem, the work tasks may be divided according to disciplines as follows: architectural design, structural design, mechanical services, plumbing and heating, transportation systems, construction, business management, and the like. Each work task is controlled by specifications to assure that the goals and objectives for the total HRB system can be at least partially attained. Indeed, each of these tasks usually involve many subtasks. For example, although the overall supervision of the construction may be conducted by a firm of consultant engineers whose sole responsibility is to serve the owners, the actual construction may be performed by a large group of subcontractors for: excavation and foundation, erection of steel frame, exterior walls, interior decoration, and so forth. As the optimization processes become fragmented in large projects, the efficiency of project organization and management become vitally important to the success of the project.

The problems and methodologies associated with project management will be discussed in Chapter 10.

4.6 METHODS OF OPTIMIZATION

The optimization methods that may be utilized for a given problem depend on the degree of order that can be established in the problem. The order reflects how well the system structure and behavior characteristics are known, and how well the system constraints and criteria can be quantified. The optimization methods can be broadly grouped into three approaches:

1. Analytical,
2. Combinatorial, and
3. Subjective.

Analytical Approach

This approach is applicable only to a totally ordered system problem. A total order implies complete knowledge of structure and behavior and either completely automates the optimization process or enables both the problem structure and the criterion preference function to be analytically stated. In such a case, the optimization process can be supported by a well-defined set of criteria that may be used to judge the quality of the design. These criteria should be included in the statements of systems objectives and should define the technical, political, and operational goals of the system.

The system variables are quantified and each variable is assigned a system value according to a common measure, such as a monetary value. Thus, if x_1, x_2, x_3, ... , and x_n are n design variables in the problem and a_i is the value of one unit of x_i to the total system, then the total value of the system may be expressed in a mathematical function as follows:

$$C = \sum_{i=1}^{n} a_i x_i \qquad (4.1)$$

The objective of optimization, then, is to find the set of x_i's such that the total system value C is maximized or minimized.

Equation 4.1 represents the simplest form of a criteria function. It is a linear equation involving only first degree terms of the design variables.

In general cases, the total system value C may appear in any functional form depending on the nature of the problem; i.e.,

$$C = F(x_1, x_2, x_3, \ldots, x_n) \qquad (4.2)$$

There may also be more than one criteria function for a given system problem. Thus, if C_1, C_2, and C_3 are three independent measures of a system, then

$$C_1 = F_1(x_1, x_2, x_3, \ldots, x_n)$$
$$C_2 = F_2(x_1, x_2, x_3, \ldots, x_n)$$
$$C_3 = F_3(x_1, x_2, x_3, \ldots, x_n)$$

And the system design must be optimized with respect to all three criteria functions. If it is not possible to optimize simultaneously all three criteria functions, then an order of priorities must be specified to resolve the conflicts.

Thus, the optimization model for a totally ordered system problem takes the following general format:

Maximize or minimize the following criteria functions:

$$C_1 = F_1(x_1, x_2, x_3, \ldots, x_n)$$
$$C_2 = F_2(x_1, x_2, x_3, \ldots, x_n)$$
$$\cdot \qquad \cdot$$
$$\cdot \qquad \cdot$$
$$\cdot \qquad \cdot$$
$$C_k = F_k(x_1, x_2, x_3, \ldots, x_n)$$

subject to the following system constraints:

Lower bounds:

$$G_1(x_1, x_2, x_3, \ldots, x_n) \geq a_1$$
$$G_2(x_1, x_2, x_3, \ldots, x_n) \geq a_2$$
$$\cdot$$
$$\cdot \qquad\qquad\qquad\qquad (4.3a)$$
$$\cdot$$
$$G_m(x_1, x_2, x_3, \ldots, x_n) \geq a_m$$

Upper bounds:

$$H_1(x_1, x_2, x_3, \ldots, x_n) \leq b_1$$
$$H_2(x_1, x_2, x_3, \ldots, x_n) \leq b_2$$
$$\cdot$$
$$\cdot \qquad\qquad\qquad\qquad (4.3b)$$
$$\cdot$$
$$H_r(x_1, x_2, x_3, \ldots, x_n) \leq b_r$$

Equalities:

$$P_1(x_1, x_2, x_3, \ldots, x_n) = c_1$$
$$P_2(x_1, x_2, x_3, \ldots, x_n) = c_2$$
$$\cdot$$
$$\cdot \qquad\qquad\qquad\qquad (4.3c)$$
$$\cdot$$
$$P_s(x_1, x_2, x_3, \ldots, x_n) = c_s$$

Approximations:

$$Q_1(x_1, x_2, x_3, \ldots, x_n) \simeq d_1$$
$$Q_2(x_1, x_2, x_3, \ldots, x_n) \simeq d_2$$
$$\cdot$$
$$\cdot \qquad\qquad\qquad\qquad (4.3d)$$
$$\cdot$$
$$Q_t(x_1, x_2, x_3, \ldots, x_n) \simeq d_t$$

Such an analytical approach is called mathematical programming. The solution of this system of mathematical expressions yields the optimum set of values for the design variables. The application of this method has already been illustrated in Section 3.5 for a road network improvement problem and will be discussed further in Chapter 5.

In some cases, the problem structure can best be described using a graphical concept, while still keeping an analytical form for the criteria preference function. The criteria function is then used to process and evaluate the relevant graphical properties exposed by the problem.

This approach is illustrated in Chapters 6 and 7.

Combinatorial Approach

As the degree of order possessed in the problem is weakened, the optimization process may reduce to purely evaluating a multiple selection of alternatives and selecting the best. Suppose that there are n design variables in the problem and that they are denoted as x_1, x_2, x_3, $\ldots$, and x_n. The approach is to identify the most probable range of values for each of these variables and then analyze all the possible combinations. If there are ten discrete probable values for each parameter, a complete combinatorial analyses will involve 10^n cases.

It is obvious that a strict application of such an approach becomes impracticable even for simple systems involving only a few variables. For example, for $n = 4$, there will already be about 10^4 cases for analysis. However, technical procedures together with past experience often can be used to eliminate a large percentage of the feasible combinations and leave only a few for detailed study and analysis.

This approach is used extensively in decision analysis. All the possibilities are analyzed and compared to arrive at an optimum decision policy. The graphical methods of paths and cut-sets analysis used in Section 2.6 for analyzing the road network improvement problem is an example of this approach. Chapters 5, 6, 7, 8, and 11 will also illustrate to varying degrees this combinatorial approach to optimization.

Subjective Approach

In complex problem situations, it may be difficult or even impossible to establish specific models or system orders, in which case purely subjective methods become necessary. The subjective approach is the most important and often is the deciding method in optimization. Intangible factors such as social values, political influences, and psychological effects are extremely difficult to quantify and measure. Yet, as all practicing engineers soon learn, these factors often control the acceptance or rejection of a design.

For example, a safe highway cannot be designed without considering the response characteristics and behavior of the drivers. Nor is it possible to find an optimum location for a reservoir without paying due regard to the political forces and ecological and social consequences involved. Whether it be in choosing an alternative design or in predicting the present and future occurrence of uncertain events, sound engineering judgments based on past experience and foresight play an indispensable role in optimization. Chapters 10 and 12 introduce the reader to areas and problem situations requiring subjective approaches to problem optimization and decision-making. In some cases the subjective elements can be quantified and incorporated in an analytical or combinatorial approach.

4.7 PROBLEMS

P4.1. Discuss the optimization concepts implicit in the following problems:

1. The total cost of a building project is the sum of the project costs (labor, material, etc.) and the indirect costs (organizational overhead, profit, etc.). Indirect costs, $\$_I$, amount to \$4,000 per project day. Direct costs, $\$_D$, are a function of the rate of use of applied resources and vary between \$100,000 for a project duration T of 100 days to about \$200,000 for a duration of 70 days and can be expressed as

$$\$_D = \frac{\$10^9}{T^2}$$

where T is the project duration in days.
Determine the project duration with minimum total cost.

2. A manufacturing company is making household appliances and its profit is a function of the number of units produced. The total profit, $\$_P$, can be expressed as

$$\$_P = 20x - 10^{-6}x^2$$

where x is the number of units produced. This expression can be interpreted to mean that profit will increase as the number of units produced increases until the market becomes saturated, and then profit will decrease. Determine the number of units to produce such that the company maximizes its profit.

3. The company in (2) has only enough plant capacity to produce 8 million units. How does this constraint affect the results obtained in (2)?

4. A City Council has just opened the contractor's bids on the construction of a new city parking lot. The bids are in terms of

unit prices; however, the entire contract will be awarded for the bid with the lowest total cost. The contractor's unit price bids and quantities required for each item are as follows:

	CLEARING	SUBBASE CU. YD.	PAVING SQ. YD.	INSTALLATION OF METERS
Contractor A	$3,000	$4.25	$26.	$25 ea.
Contractor B	4,000	4.00	25.	30 ea.
Contractor C	2,000	3.75	24.	35 ea.
Required	1	1×10^4	7×10^4	500

Which contract should the City Council accept?

P4.2. Public facilities (such as fire stations) have many different characteristics that contribute to their value or acceptance by society and prevent the formulation of unique criteria for selecting optimal facility designs. In these cases it may be possible to establish no more than a subjective preference for one complex design over another in a series of binary decisions. This situation leads to the development of preference graphs (see Kaufman, 1968) in which each facility design as it incorporates more features is modeled and labeled by additional design space nodes. Preference for one design over another is then modeled by a directed arrow from the less preferable to the more preferable facility.

What factors do you think should be considered in locating and designing a city fire station?

Develop a preference graph for a number of fire station situations that have various levels of desirable features.

P4.3. For the hierarchical systems that were defined in P1.3, define the goals, objectives, and possible criteria for each component and system level. As the system levels increase in scope, why is it more difficult to reach a truly optimum solution? How does this situation affect the optimization method or technique that is utilized?

P4.4. The absolute and relative locations of physical facilities in a production plant are simple ways of describing or modeling the plant. How good or efficient the plant layout becomes depends on the goals, objectives, and criteria that are used to evaluate the plant.

1. Structure the following goals and objectives relevant to a particular plant you are familiar with:

minimize product handling
minimize inventory levels within the process line
minimize floor space
minimize cost
minimize work force
maximize mobility to change process lines
maximize use of existing capital equipment
maximize profit
maximize output

2. The investigation and modeling of a given plant will focus on a specific set of plant characteristics. The characteristics selected and hence the modeling abstraction are directly related to the goal, objective, and criteria under consideration.

Select two different objectives and show how different modeling rationales develop.

P4.5. The cost of environmental services such as heating, air conditioning, lights, etc., may constitute as much as one-half the cost of a building. The following represents various levels of optimization of the cost of the environmental services of a high-rise building:

1. The number of floors, the amount of floor space, the use of the building, and the building material are specified and the objective is to minimize the cost of the environmental services.

2. The number of floors, the amount of floor space, and the use of the building are specified and the objective is to minimize the cost of the building material and environmental services.

3. The amount of floor space and use is specified and the objective is to minimize the cost of the building.

4. The owner wants to invest in a high-rise building that will allow him to maximize his profit and he doesn't care how large it is or what it is used for.

Describe the problems that arise at each level of optimization. How could trade-offs be developed at each level of optimization? Explain why you think it is easier or more difficult to optimize as the constraints are removed.

P4.6. Large-scale water resource projects are usually public investment type developments in that they are built with public money for public use and benefit.

1. What might be some goals and objectives for public investment in water resources development? What criteria might you use for measuring the effectiveness of the projects in achieving the objectives you listed? How are goals and objectives for public investment determined?

2. As part of the water resources development of a region, a decision has been made to build a water supply reservoir. What might be some of the objectives that the engineer would try to meet in designing the reservoir? How are these related to the goals and objectives listed in (1)? What criteria might be used? How might trade-offs be developed among conflicting objectives? (Hint: See Hall and Dracup, 1970).

P4.7. A river bisects a large city and is crossed by a number of bridges in the city transportation system. The city engineer is asked to locate a new and additional bridge over the river in a section between two existing bridges.

What criteria may develop that the engineer can use to evaluate a specific bridge location and its influence on the city transportation system?

What political, social, economical, and technical constraints might affect the bridge location?

P4.8. In many design and planning situations the decision-making and optimization processes are hampered by the difficulty and delay in formulating system goals and objectives (see Hitch, 1961).

Discuss the above statement in reference to a local problem you are familiar with.

5

MATHEMATICAL PROGRAMMING

*Armco Steel Corporation's $360 million dollar steel making and rolling mill complex.
The Armco complex includes a $39 million advanced pollution abatement system.
The Armco complex was selected as the Outstanding Civil Engineering Achievement
of 1970 for construction innovations, techniques, and materials and for leadership in
seeking solutions to industrial pollution problems. (Courtesy of the American Society
of Civil Engineers)*

5.1   THE NATURE OF MATHEMATICAL PROGRAMMING

Mathematical programming is used to find the best or optimal solution to a problem that requires a decision or set of decisions about how best to use a set of limited resources to achieve a stated goal or objective. It makes three important contributions to the study of the problem.

1. The real-life situation must be structured into a mathematical model that abstracts the essential elements so that a solution relevant to the decision-maker's objective can be sought. This involves looking at the problem in the context of the entire system.

2. The structure of solutions must be explored and systematic procedures must be developed for obtaining the solutions.

3. It yields an optimal value of desirability for the system or has at least compared alternative courses of action by evaluating their desirability.

Mathematical programming is discussed in this chapter in terms of linear programming and nonlinear programming. Dynamic programming is discussed in Chapter 11.

5.2   THE LINEAR PROGRAMMING MODEL

Linear programming, the simplest and most widely used of the programming techniques, requires that all the mathematical functions in the model be linear functions.

The mathematical statement of a general linear programming problem is the following: Find the values $x_1, x_2, x_3, \ldots, x_n$ (called decision variables) that maximize the linear function Z (i.e., the criterion function, commonly called the objective function)

$$Z = c_1 x_1 + c_2 x_2 + \ldots + c_n x_n \qquad (5.1)$$

subject to the following relationships (called constraints or restrictions):

$$a_{11}x_1 + a_{12}x_2 + \ldots + a_{1n}x_n \leq b_1$$
$$a_{21}x_1 + a_{22}x_2 + \ldots + a_{2n}x_n \leq b_2$$

$$\cdot \qquad\qquad \cdot$$

$$\cdot \qquad\qquad \cdot \qquad\qquad (5.2)$$

$$\cdot \qquad\qquad \cdot$$

$$a_{m1}x_1 + a_{m2}x_2 + \ldots + a_{mn}x_n \leq b_n$$

$$\text{all } x_j \geq 0$$

where

$$a_{ij}, b_i, \text{ and } c_j = \text{given constants.}$$

The linear programming model can be written in more efficient notation as

Maximize

$$Z = \sum_{j=1}^{n} c_j x_j$$

subject to:

$$\sum_{j=1}^{n} a_{ij} x_j \leq b_i$$

where   (5.3)

$$i = 1, 2, \ldots, m$$

and

$$x_j \geq 0$$

where

$$j = 1, 2, \ldots, n$$

The decision variables, $x_1, x_2, \ldots, x_n$, represent levels of n competing activities. These decision variables represent variables in the real situation being modeled that can be freely changed in magnitude by management. If each activity is manufacturing a certain product, then x_j would be the number of units of the jth product to be produced during a given time period; Z is the overall measure of effectiveness (e.g., profit over the given time period); and c_j is the increase in the objective function (profit) that would result from each unit increase in x_j. Each of the first m linear inequalities corresponds to a restriction on the availability of one of the resources; b_i is the total amount of the ith resource available; and a_{ij} is the amount of the ith resource used for each unit of the jth product. The restriction that all x_j be greater than or equal to zero models the fact that negative quantities cannot be produced. This constraint implies that the minimum value that any decision variable can attain is zero. The "less than or equal" sign in the constraint inequalities may be replaced by "greater than or equal," "equal," or "approximately equal" signs to suit the description of the particular problem that is being modeled.

5.3 DEVELOPING LINEAR PROGRAMMING MODELS

The variety of situations to which linear programming has been applied ranges from agriculture to zinc smelting. However, it is not always easy to recognize that a decision-making problem can be solved by linear programming. A certain amount of skillful trial and error is usually required to capture the essence of a decision problem in a linear model. The decision-maker must determine the objective that he wishes to consider and this objective must be described by a criterion function in terms of the decision variables, x_i, about which he must make a decision or which he can control. The decision variables are not always obvious at first glance. Furthermore, the decision-maker is always limited by constraints on what he can do. He must exercise care in depicting the relevant constraints to be imposed on the optimization. An analysis procedure is then followed, which leads to the selection of values for the decision variables that optimize the criterion function while satisfying all the constraints imposed on the problem. One way to gain an understanding of the types of problems to which linear programming can be applied is to study how linear models have been formulated for several problems that have been presented verbally—the form in which the decision-maker usually meets the problem.

A Product Mix Problem

General Problem A manufacturer has fixed amounts of different resources such as raw material, labor, and equipment. These resources can be combined to produce any one of several different products. He knows how much of the ith resource it takes to produce one unit of the jth product. He also knows how much profit he makes for each unit of the jth product. He wishes to produce the combination of products that will maximize total profit.

Example Problem The N. Dustrious Company produces two products: I and II. The raw material requirements, space needed for storage, production rates, and profits for these products are given in Table 5.1. The total amount of raw material available per day for both products is 1,575 lb, the total storage space for all products is 1,500 sq ft,

Table 5.1

PRODUCTION DATA FOR N. DUSTRIOUS COMPANY

	PRODUCT	
	I	II
Storage space (sq ft/unit)	4	5
Raw material (lbs/unit)	5	3
Production rate (units/hr)	60	30
Profit ($/unit)	13	11

and a maximum of 7 hr/day can be used for production. All products manufactured are shipped out of the storage area at the end of the day. Therefore, the two products must share the total raw material, storage space, and production time. The company wants to determine how many units of each product to produce per day in order to maximize its total profit.

Formulating the Problem   The company has decided that it wants to maximize its profit, which depends on the number of units of product I and II that it produces. Therefore, the decision variables, x_1 and x_2 can be the number of units of products I and II, respectively, produced per day. The object, then, is to maximize the equation

$$Z = 13x_1 + 11x_2$$

subject to the constraints on storage space, raw materials, and production time. Each unit of product I requires 4 sq ft of storage space and each unit of product II requires 5 sq ft. Thus, a total of $4x_1 + 5x_2$ sq ft of storage space is needed each day. This space must be less than or equal to the available storage space, which is 1,500 sq ft. Therefore,

$$4x_1 + 5x_2 \leq 1,500$$

Similarly, each unit of product I and II produced requires 5 and 3 lbs, respectively, of raw material. Hence, a total of $5x_1 + 3x_2$ lbs of raw material is used. This must be less than or equal to the total amount of raw material available, which is 1,575 lbs. Therefore,

$$5x_1 + 3x_2 \leq 1,575$$

The limitation on production time is a little more difficult to translate into a mathematical statement. Product I can be produced at the rate of 60 units per hour. Therefore, it must take 1 minute or 1/60 of an hour to produce one unit. Similarly, it can be reasoned that it requires 1/30 of an hour to produce one unit of product II. Hence, a total of $x_1/60 + x_2/30$ hours is required for the daily production. This quantity must be less than or equal to the total production time available each day. Therefore,

$$\frac{x_1}{60} + \frac{x_2}{30} \leq 7$$

This expression can be cleared of fractions by multiplying each term by 60. The result is

$$x_1 + 2x_2 \leq 420$$

Finally, the company cannot produce a negative quantity of any product (it is impossible to produce less than nothing). This means that x_1 and x_2 must each be greater than or equal to zero.

The linear programming model for this example can be summarized as:

Maximize
$$Z = 13x_1 + 11x_2$$
subject to:
$$4x_1 + 5x_2 \leq 1{,}500$$
$$5x_1 + 3x_2 \leq 1{,}575$$
$$x_1 + 2x_2 \leq 420 \tag{5.4}$$
$$x_1 \geq 0$$
$$x_2 \geq 0$$

The above model is typical of most linear programming models. A specific objective (i.e., profit) has been chosen from the many possible objectives that may pertain to the situation; i.e., cost of production, full use of machines. In addition, the number of constraint statements that can be formulated is not rigorously prescribed, but depends on the perception of the modeler of features in the real problem. This emphasizes the creative aspects of linear programming modeling.

A Blending Problem

General Problem Blending problems refer to situations in which a number of components (or commodities) are mixed together to yield one or more products. Typically, different commodities are to be purchased. Each commodity has known characteristics and costs. The problem is to determine how much of each commodity should be purchased and blended with the rest so that the characteristics of the mixture lie within specified bounds and the total cost is minimized.

Example Problem The Sunnyflush Company has two plants located along a stream, as shown in Figure 5.1. Plant 1 is generating 20 units of pollutants daily and plant 2 is generating 14 units. Before the wastes are discharged into the stream, part of these pollutants are removed by a waste treatment facility in each plant. The costs associated with removing a unit of pollutant are $1,000 and $800 for plants 1 and 2, respectively. The rates of flow in the streams are Q_1 equal to 5 million gallons per day (mgd) and Q_2 equal to 2 mgd, and the flows contain no pollutants until they pass the plants. Stream standards require that the number of units of pollutants per million gallons of flow should not exceed 2. Twenty percent of the pollutants entering the stream at plant 1

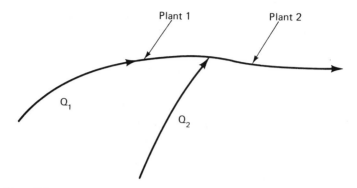

Figure 5.1

Location of Sunnyflush Company's Plants

will be removed by natural processes (i.e., oxidation by the air) before they reach plant 2. The Sunnyflush Company wants to determine the most economical operation of its waste treatment facilities that will allow it to satisfy the stream standards.

Formulating the Problem   The most economical operation of the waste treatment plants would be the operation that minimizes the total cost of removing part of the pollutants from the wastes of both plants. The decisions to be made, then, are how many units of pollutants are to be removed by each waste treatment facility. Let x_1 and x_2 be the number of units of pollutants to be removed by the waste treatment facilities of plants 1 and 2, respectively. The total cost of removing the pollutants is $1{,}000x_1 + 800x_2$ and is the quantity to be minimized. If x_1 units of pollutants are removed from the wastes at plant 1, then $20 - x_1$ units of pollutants are released into the stream. This quantity cannot be greater than the quantity of pollutants permitted by the stream standards, which is 2 times the flow rate Q_1; i.e., 10 units. Thus, the first constraint is:

$$20 - x_1 \leq 10$$

The second point of concern is at plant 2. Here the amount of pollutants released is $14 - x_2$ units. In addition, at the outlet from plant 2 there are pollutants from plant 1. Since 20 percent of the pollutants released at plant 1 are removed by natural processes, 80 percent remains by the time the flow reaches plant 2. Therefore, the total amount of pollutants in the stream at plant 2 is $.8(20 - x_1) + (14 - x_2)$. This quantity must not be greater than the stream standards allow. Thus, the second constraint is:

$$0.8(20 - x_1) + (14 - x_2) \leq 14$$

where the right-hand side, 14, is 2 times the combined flow rate of Q_1 and Q_2. Since the waste treatment facilities cannot remove a negative amount, x_1 and x_2 must each be zero or positive. Furthermore, the treatment facilities cannot remove more pollutants than exist in the waste, which means that x_1 must be less than or equal to 20 and x_2 must be less than or equal to 14. The linear programming model for this example can be summarized as:

Minimize

$$Z = 1,000x_1 + 800x_2$$

subject to:

$$20 - x_1 \leq 10$$
$$0.8(20 - x_1) + (14 - x_2) \leq 14$$
$$x_1 \leq 20$$
$$x_2 \leq 14 \tag{5.5}$$
$$x_1 \geq 0$$
$$x_2 \geq 0$$

The constraints can further be simplified to:

$$0.8x_1 + x_2 \geq 16$$
$$20 \geq x_1 \geq 10 \tag{5.6}$$
$$14 \geq x_2 \geq 0$$

A Production Scheduling Problem

General Problem   A manufacturer knows that he must supply a given number of items of a certain product each month for the next n months. They can be produced either in regular time, subject to a maximum each month, or in overtime. The cost of producing an item during overtime is greater than during regular time. A storage cost is associated with each item not sold at the end of the month. The problem is to determine the production schedule that minimizes the sum of production and storage costs.

Example Problem   The Hard Rock Company provides gravel from its quarry for concrete mixing companies. During the next 4 months its sales, costs, and available time are expected to be those given in

Table 5.2. There is no gravel in stock at the beginning of the first month. It takes 1.5 hours of production time to produce 1,000 cubic yards (cu yds) of gravel. It costs $5 to store 1,000 cu yds of gravel from one month to the next. The company wants a production schedule that does not exceed the production time limitations each month, that meets the demand requirements, and that minimizes total cost.

Table 5.2

SALES, COSTS, AND AVAILABLE TIME FOR HARD ROCK COMPANY

	MONTH			
	1	2	3	4
Gravel required (1,000 cu yds)	1,000	2,500	2,100	2,900
Cost regular time ($/1,000 cu yds)	100	100	110	110
Cost overtime ($/1,000 cu yds)	110	116	120	124
Regular operation time of rock crushers, (hrs)	2,400	2,400	2,400	2,400
Overtime (hrs)	990	990	990	990

Formulating the Problem   Let x_1, x_2, x_3, and x_4 be the number of units of gravel produced in months 1, 2, 3, and 4, respectively, on regular time where a unit of gravel is 1,000 cu yds. Also, let x_5, x_6, x_7, and x_8 be the number of units of gravel produced during months 1, 2, 3, and 4, respectively, on overtime, and let x_9, x_{10}, and x_{11} be the number of units of gravel in stock at the end of months 1, 2, and 3, respectively. This assumes that there is to be no gravel left in stock at the end of month 4. The objective is to minimize costs. Therefore, the company wants to minimize

$$Z = 100x_1 + 100x_2 + 110x_3 + 110x_4 + 110x_5 + 116x_6$$
$$+ 120x_7 + 124x_8 + 5x_9 + 5x_{10} + 5x_{11}$$

subject to demand requirements for gravel sales and production time limitations. In order to meet sales demands in the first month, the total number of units of gravel produced ($x_1 + x_5$) must be equal to the sum of the sales demand (1,000 cu yds) and the number of units in storage at the end of the month, x_9. Thus, the first constraint is

$$x_1 + x_5 = 1,000 + x_9$$

During the second month, the quantity in stock at the beginning of the month, x_9, plus the quantity produced, $x_2 + x_6$, must be equal to the sum of the quantity on stock at the end of the month, x_{10}, and the sales demand of 2,500 units. Hence, the second constraint is

$$x_9 + x_2 + x_6 = 2{,}500 + x_{10}$$

Similarly, the third and fourth constraints are

$$x_{10} + x_3 + x_7 = 2{,}100 + x_{11}$$

and

$$x_{11} + x_4 + x_8 = 2{,}900$$

In addition, the production time limitations cannot be violated. Therefore,

$$1.5x_i \leq 2{,}400 \text{ where } i = 1, 2, 3, \text{ and } 4$$
$$1.5x_i \leq 990 \quad \text{where } i = 5, 6, 7, \text{ and } 8$$

Finally, a negative quantity of gravel cannot be produced at any time, and thus each x_i must be greater than or equal to zero.

The linear programming model for this problem can be summarized as

Minimize

$$Z = 100x_1 + 100x_2 + 110x_3 + 110x_4 + 110x_5 + 116x_6$$
$$+ 120x_7 + 124x_8 + 5x_9 + 5x_{10} + 5x_{11}$$

subject to:

$$x_1 + x_5 - x_9 = 1{,}000$$
$$x_9 + x_2 + x_6 - x_{10} = 2{,}500$$
$$x_{10} + x_3 + x_7 - x_{11} = 2{,}100 \tag{5.7}$$
$$x_{11} + x_4 + x_8 = 2{,}900$$
$$1.5x_i \leq 2{,}400 \text{ where } i = 1, 2, 3, \text{ and } 4$$
$$1.5x_i \leq 990 \quad \text{where } i = 5, 6, 7, \text{ and } 8$$
$$x_i \geq 0 \quad \text{where } i = 1, 2, \ldots, 11$$

A linear graph model can be developed to illustrate the derivation of the constraint operation. In Figure 5.2, for example, the nodes represent months and the branches storage stocks. Nodal inputs and outputs model crusher production and sales. Continuity statements at nodes and capacity conditions produce the system constraints. A formal mathematical solution to Equation 5.7 is possible, as indicated later, but the reader may develop a graphical solution based on Figure 5.2.

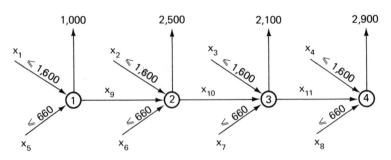

Figure 5.2

Gravel Flow Model

A Transportation Problem

General Problem   A product is to be shipped in the amounts a_1, a_2, ... , a_m from m shipping origins and received in amounts b_1, b_2, ... , b_n at each of n shipping destinations. The cost of shipping a unit from the ith origin to the jth destination is known for all combinations of origins and destinations. The problem is to determine the amount to be shipped from each origin to each destination such that the total cost of transportation is a minimum.

Example Problem   The Cheap Concrete Company has three sand pits from which it can obtain sand for its two concrete plants. Plant 1 requires 24 units of sand, and Plant 2 requires 28 units. Unfortunately, the sand pits are almost depleted. Pit 1 has only 8 units of sand left, pit 2 has 23 units, and pit 3 has 21 units. It costs $19 to haul 1 unit of sand from pit 1 to plant 1. The remaining haul costs are given in Table 5.3. The company wants to determine how much to haul from each pit to each plant in order to minimize its total hauling cost.

Table 5.3

HAULING COSTS

	PLANT	
PIT	1	2
1	$19	$12
2	6	9
3	7	18

Formulating the Problem   Let $x_{i,j}$ be the number of units of sand hauled from the ith pit to the jth plant. To minimize hauling costs, the objective function is

Minimize

$$Z = 19x_{1,1} + 12x_{1,2} + 6x_{2,1} + 9x_{2,2} + 7x_{3,1} + 18x_{3,2} \qquad (5.8)$$

subject to availability and demand constraints. The quantity hauled from pit 1, $x_{1,1} + x_{1,2}$, cannot be greater than the amount available at pit 1; i.e.,

$$x_{1,1} + x_{1,2} \leq 8$$

similarly, for pits 2 and 3

$$x_{2,1} + x_{2,2} \leq 23$$
$$x_{3,1} + x_{3,2} \leq 21 \qquad (5.9)$$

The total amount hauled from the three pits to plant 1; i.e., $x_{1,1} + x_{2,1} + x_{3,1}$, must equal the quantity needed at plant 1; i.e.,

$$x_{1,1} + x_{2,1} + x_{3,1} = 24$$

similarly, at plant 2 $\qquad\qquad\qquad\qquad\qquad\qquad\qquad\qquad (5.10)$

$$x_{1,2} + x_{2,2} + x_{3,2} = 28$$

Finally, each $x_{i,j}$ must be positive or zero since it is impossible to haul a negative quantity.

Special cases of the transportation problem include the transshipment and the assignment problems. Transshipment can occur in the distribution system of a company that has regional warehouses that ship to smaller district warehouses, which in turn ship to final destinations. The transshipment problem is a direct extension of the transportation problem and permits each source or destination to act as an intermediate point for shipments from other sources to other destinations. The assignment problem occurs when n individuals or machines must be assigned to perform n different jobs. Each individual has a rating of effectiveness on each job and the objective is to maximize the total effectiveness for all n jobs.

Although the transportation, transshipment, and assignment problems can be formulated as linear programming models, more efficient algorithms exist for solving these problems (Dantzig, 1963; Gass, 1969; Lleywellyn, 1964; Wagner, 1969).

A Flow-Capacity Problem

The mathematical models developed in Section 3.5 for the flow capacity problem described in Section 2.4 are linear programming models.

The models in Section 3.5 illustrate that different linear programming models can be developed for the same system depending on the objective of the decision-maker and the constraints that must be satisfied. For most problems, once the model has been formulated, there are algorithms for solving the model to determine the values for the unknown variables. One of these algorithms is explained below.

5.4 GRAPHICAL SOLUTION TO LINEAR PROGRAMMING PROBLEM

In this day of the high-speed digital computer, the graphical method of solution is still useful for small problems involving two decision variables, and it also provides clues to a systematic method of solving linear programming problems that have more than two decision variables.

Consider the following linear programming model that was formulated above as a product mix problem.

Maximize
$$Z = 13x_1 + 11x_2$$
subject to:
$$4x_1 + 5x_2 \leq 1,500$$
$$5x_1 + 3x_2 \leq 1,575 \qquad (5.11)$$
$$x_1 + 2x_2 \leq 420$$
$$x_1 \geq 0; \qquad x_2 \geq 0$$

Any pair of values of x_1 and x_2 that satisfies the set of inequalities is a feasible solution to the problem. An optimal solution is a feasible solution that also maximizes the objective function.

Since the problem consists of only two variables, x_1 and x_2, a graphical method may be used to obtain the solution. An equation of the form $4x_1 + 5x_2 = 1,500$ defines a straight line in the x_1, x_2 plane. An inequality defines an area bounded by a straight line. Therefore, the region below and including the line $4x_1 + 5x_2 = 1,500$ in Figure 5.3 represents the region defined by $4x_1 + 5x_2 \leq 1,500$. The region below and including the line $5x_1 + 3x_2 = 1,575$ in Figure 5.3 represents the region defined by $5x_1 + 3x_2 \leq 1,575$. Similarly, the region defined by $x_1 + 2x_2 \leq 420$ is represented in Figure 5.3 by the region below and including the line $x_1 + 2x_2 = 420$. The shaded area of Figure 5.3 comprises the area common to all the regions defined by the constraints and contains all pairs of x_1 and x_2 that are feasible solutions to the problem. This area is known as the feasible region or feasible solution space. The optimal solution must lie within this region.

There are various pairs of x_1 and x_2 that satisfy the constraints.

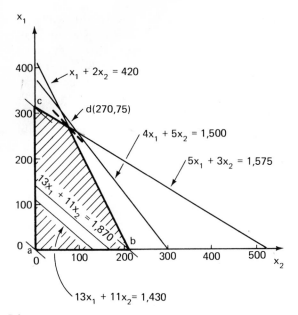

Figure 5.3

Graphical Solution for Linear Programming Problem

One pair, $x_1 = 0$ and $x_2 = 0$, satisfies the constraints and provides a profit equal to 0. The pair of variables in a solution can be written in vector form.

$$\mathbf{X} = \begin{bmatrix} x_1 \\ x_2 \end{bmatrix} = \begin{bmatrix} 0 \\ 0 \end{bmatrix}$$

Various pairs of x_1 and x_2 lead to the same profit. A profit of \$1,430, for instance, would be obtained for all pairs of values of x_1 and x_2 satisfying the equation $13x_1 + 11x_2 = 1,430$; i.e., lying on the line representing this equation in Figure 5.3. The line for any other value of the profit would be a line parallel to $13x_1 + 11x_2 = 1,430$, such as $13x_1 + 11x_2 = 1,870$. The line representing profit would be further from the origin for a larger profit, but would be nearer the origin for a smaller profit. The maximum value of the objective function that satisfies all the constraints is determined by finding the point that, although within the feasible region, lies on a line parallel to the line $13x_1 + 11x_2 = 1,430$ and is as far from the origin as possible. From Figure 5.3 it can be seen that point d is such a point.

The optimal solution from Figure 5.3 is $\mathbf{X} = \begin{bmatrix} 270 \\ 75 \end{bmatrix}$ and means that a maximum profit of \$4,335 is obtained by producing 270 units of product I and 75 units of product II. In this solution, all the raw material and available time are used, because the optimal point lies on the two constraint lines for these resources. However, $1{,}500 - [4(270) + 5(75)]$, or 45 sq ft of storage space, is not used. Thus, the storage space is not a constraint on the optimal solution; i.e., more products could be produced before the company ran out of storage space. Thus, this constraint is said to be slack.

The graphical concepts associated with the feasible space of a problem gives some insight into why linear programming problems are generally over- or under-determined sets of inequations (i.e., where the number of variables are more than or less than the number of inequations). In this case a constraint inequation is produced for the three relationships that can be determined between x_1 and x_2; two of these effectively assist in defining the feasible space, one is redundant.

It may be that a problem does not have a feasible solution space. If the above problem had another constraint such that $2x_1 + x_2 \geq 800$, there would be no pair of x_1 and x_2, which would satisfy all the constraints.

If there is a feasible region for a linear programming problem, it may be represented graphically by a polygon. Notice in Figure 5.3, that a linear objective function must attain its maximum value at an extreme point of such a region. The feasible region of Figure 5.3 is convex; i.e., all internal angles are less than 180 degrees. The graphical method of solution can also be applied to nonconvex regions; i.e., regions which have at least one interior angle that is greater than 180 degrees, but the simplex method described below is only applicable to convex regions.

If the objective function happens to be parallel to one of the edges of the feasible region, then any point along this edge between the two extreme points may be an optimal solution that maximizes the objective function. When this occurs, there is no unique solution, but there is an infinite number of optimal solutions.

The graphical method of solution may be extended to a case in which there are three variables. In this case, each constraint is represented by a plane in three dimensions, and the feasible region bounded by these planes is a polyhedron. As in the two-variable case, the objective function attains its maximum value at an extreme point of the three-dimensional feasible region. Although the graphical method of solution is of no practical value when the number of decision variables exceeds three, it does provide clues to the simplex method.

5.5 THE SIMPLEX METHOD

The simplex method is not used to examine all the feasible solutions. It deals only with a small and unique set of feasible solutions, the set of vertex points (i.e., extreme points) of the convex feasible space that contains the optimal solution. It considers the members of this set one at a time, and ends when the optimum has been achieved. The simplex method involves the following steps:

1. Locate an extreme point of the feasible region.

2. Examine each boundary edge intersecting at this point to see whether movement along any edge increases the value of the objective function.

3. If the value of the objective function increases along any edge, move along this edge to the adjacent extreme point. If several edges indicate improvement, then the edge providing the greatest rate of increase is selected.

4. Repeat steps 2 and 3 until movement along any edge no longer increases the value of the objective function.

The Simplex Method Applied

The simplex method is an algebraic method. Because inequality equations are less amenable to algebraic manipulation than equations, the linear programming model must first be converted into an equivalent model containing no inequalities. The simplex method assumes nonnegativity; therefore, the nonnegativity constraints can be ignored. Inequalities can be converted to equalities by adding slack variables.

Consider the inequality $4x_1 + 5x_2 \leq 1{,}500$. Let $x_3 = 1{,}500 - 4x_1 - 5x_2$ so that x_3 is the slack between the two sides of the inequality; hence, x_3 is called a slack variable. Now $4x_1 + 5x_2 + x_3 = 1{,}500$. The original constraint, $4x_1 + 5x_2 \leq 1{,}500$, holds whenever $x_3 \geq 0$. Therefore, $4x_1 + 5x_2 \leq 1{,}500$ is equivalent to, and can be replaced by, the set of constraints $4x_1 + 5x_2 + x_3 = 1{,}500$ and $x_3 \geq 0$.

Introducing slack variable for the other constraints, except the nonnegativity restrictions, and rewriting the objective function such that all variables are on the left-hand side of the equation, Equation 5.11 can be expressed as:

$$
\begin{array}{llll}
Z - 13x_1 - 11x_2 & & = 0 & \text{(A)} \\
4x_1 + 5x_2 + x_3 & & = 1{,}500 & \text{(B)} \\
5x_1 + 3x_2 & + x_4 & = 1{,}575 & \text{(C)} \\
x_1 + 2x_2 & + x_5 & = 420 & \text{(D)} \\
\end{array}
$$

$$x_i \geq 0 \qquad i = 1, 2, \ldots, 5$$

(5.12)

where the new variables, x_3, x_4, and x_5 are called slack variables. Slack variables represent the quantity of each resource that is not used. Thus, x_3 represents the amount of storage space that is not used when x_1 units of product I and x_2 units of product II are produced. The goal is now to obtain nonnegative values of x_1, x_2, x_3, x_4, and x_5 that will satisfy the three new equality constraints and maximize the value of the objective function.

The first step in the simplex method is to locate an extreme point of the feasible region. An obvious solution is to let $x_1 = 0$ and $x_2 = 0$. This implies that $x_3 = 1,500$, $x_4 = 1,575$, and $x_5 = 420$. This set of values

$$\mathbf{X} = \begin{bmatrix} 0 \\ 0 \\ 1,500 \\ 1,575 \\ 420 \end{bmatrix} \tag{5.13}$$

is called a basic feasible solution. A basic solution has exactly as many variables with nonzero values as there are constraint equations. A basic solution is a basic feasible solution if it also satisfies all the constraint equations. The nonzero valued variables (x_3, x_4, and x_5 in this step) are referred to as the basic variables or as the variables in the basis. The zero-valued variables (x_1 and x_2 in this step) are referred to as nonbasic variables. Referring back to Figure 5.3, this procedure has selected the origin as the initial extreme point (or basic feasible solution), and the value of the objective function is zero.

Given a basic feasible solution, the simplex method obtains a new basic feasible solution by selecting an adjacent extreme point that increases the value of the objective function or discovers that the present solution is optimum. The algebraic equivalent of an adjacent extreme point is a new basic feasible solution with all but one of the same variables in the basis. Thus, selecting the next basic feasible solution requires selecting one basic variable to leave the present basis and become nonbasic, and selecting one nonbasic variable to enter the basis.

The entering variable is chosen as the variable that causes the greatest rate of increase in the objective function. The objective function $Z = 13x_1 + 11x_2$ can be increased by bringing either x_1 or x_2 into two basis. Z increases at the rate of 13 per unit increase in x_1 and 11 per unit increase in x_2. Since 13 is greater than 11, x_1 is chosen as the entering variable.

Either x_3, x_4, or x_5 must be chosen as the leaving variable. Since the objective is to maximize Z, which increases as x_1 increases, the variable to leave the basis should be the one that first becomes zero as x_1 increases. This increases Z as much as is feasible by increasing the entering variable x_1. The constraint equations can be rewritten with the basic variables only on the left-hand side. Thus,

$$x_3 = 1{,}500 - 4x_1 - 5x_2$$
$$x_4 = 1{,}575 - 5x_1 - 3x_2 \qquad (5.14)$$
$$x_5 = 420 - x_1 - 2x_2$$

If $x_2 = 0$, $x_3 = 0$ when $x_1 = 375$, $x_4 = 0$ when $x_1 = 315$, and $x_5 = 0$ when $x_1 = 420$. Therefore, the maximum increase in x_1 is 315 and x_4 is the variable to leave the basis. This is the maximum value that x_1 can attain in the new basis, because if x_1 is greater than 315 the value of x_4 becomes negative, which would violate the nonnegativity constraint.

The next step is to solve for the new values of the remaining basic variables. This is done by algebraically manipulating the equations until Z and the basic variables can be expressed in terms of only the nonbasic variables. This requires that Z and each basic variable appear in exactly one equation and that this equation contain no other basic variable. The variable Z is considered as a basic variable while operating on the objective function. To achieve the new equations,

1. Divide Equation C by 5 to obtain Equation C1;
2. Multiply Equation C by $-4/5$ and add to Equation B to obtain Equation B1;
3. Multiply Equation C by $-1/5$ and add to Equation D to obtain Equation D1;
4. Multiply Equation C by $+13/5$ and add to Equation A to obtain Equation A1;

$$Z \quad - \frac{16}{5} x_2 \quad + \frac{13}{5} x_4 \qquad = 4{,}095 \qquad \text{(A1)}$$

$$\frac{13}{5} x_2 + x_3 - \frac{4}{5} x_4 \qquad = 240 \qquad \text{(B1)}$$

$$x_1 + \frac{3}{5} x_2 \quad + \frac{1}{5} x_4 \qquad = 315 \qquad \text{(C1)}$$

$$+ \frac{7}{5} x_2 \quad - \frac{1}{5} x_4 + x_5 = 105 \qquad \text{(D1)}$$

$$(5.15)$$

Since the nonbasic variables x_2 and x_4 equal 0, the new basic feasible solution is

$$X = \begin{bmatrix} 315 \\ 0 \\ 240 \\ 0 \\ 105 \end{bmatrix} \qquad (5.16)$$

with a Z value of 4,095. Referring to Figure 5.3, this corresponds to point c, which is the intersection of the x_1 axis and the line for equation $5x_1 + 3x_2 = 1,575$.

Referring to Equation A1, Z can be increased if x_2 is increased. Therefore, the optimal solution has not been obtained, so x_2 must now be brought into the basis and either x_1, x_3, or x_5 must leave the basis. Referring to Equations B1, C1, and D1, x_2 can increase to $92\frac{4}{13}$ when $x_3 = 0$, to 525 when $x_1 = 0$, and to 75 when $x_5 = 0$. Thus, the maximum x_2 can attain and not violate the nonnegativity constraints is 75, and, therefore, the leaving variable is x_5. Now, Z and the basic variables must be expressed in terms of the nonbasic variables. This can be done by algebraically manipulating Equations A1, B1, C1, and D1 similar to the procedure above to obtain

$$Z \qquad + \frac{15}{7}x_4 + \frac{16}{7}x_5 = 4,335 \qquad (A2)$$

$$x_3 - \frac{3}{7}x_4 - \frac{13}{7}x_5 = 45 \qquad (B2)$$

$$x_1 \qquad + \frac{10}{35}x_4 - \frac{3}{7}x_5 = 270 \qquad (C2)$$

$$x_2 \qquad - \frac{1}{7}x_4 + \frac{5}{7}x_5 = 75 \qquad (D2)$$

$$(5.17)$$

such that the new basic feasible solution is

$$X = \begin{bmatrix} 270 \\ 75 \\ 45 \\ 0 \\ 0 \end{bmatrix} \qquad (5.18)$$

with a Z value of 4,335.

From Equation A2 it is clear that if one of the nonbasic variables, x_4 or x_5, is increased, then the value of Z will decrease. Therefore, the optimal solution has been obtained. This corresponds to point d in Figure 5.3.

The simplex method can be summarized by use of a simplex

tableau, which is used to record only the coefficients of the variables and the constants on the right-hand side of the equations. The simplex tableau procedure for this problem is summarized in Table 5.4.

Simplex Tableau Procedure for Maximization Problem

1. Rewrite the objective function with all variables on the left-hand side of the equation, add slack variables to transform inequalities into equalities, and then arrange coefficients and constants on right-hand side of equations into the initial tableau. The slack variables are initial basic variables.

2. If the objective function now contains negative coefficients, choose the nonbasic variable with the smallest coefficient (i.e., the negative coefficient with the largest absolute value) as the entering basic variable. If two or more nonbasic variables are tied with the smallest coefficient, select one of these arbitrarily and continue.

Table 5.4

SUMMARY OF SIMPLEX TABLEAU PROCEDURE FOR
MAXIMIZATION PROBLEM

ROW NO.	BASIC VARI-ABLES	Z	x_1	x_2	x_3	x_4	x_5	RHS	UPPER BOUND ON ENTERING VARIABLE
Initial tableau									
1a	Z	1	-13	-11	0	0	0	0	
2a	x_3	0	4	5	1	0	0	1,500	375
3a	x_4	0	5	3	0	1	0	1,575	315
4a	x_5	0	1	2	0	0	1	420	420
Tableau at end of 1st iteration									
1b	Z	1	0	$-\frac{16}{5}$	0	$+\frac{13}{5}$	0	4,095	
2b	x_3	0	0	$\frac{13}{5}$	1	$-\frac{4}{5}$	0	240	92.3
3b	x_1	0	1	$\frac{3}{5}$	0	$\frac{1}{5}$	0	315	525
4b	x_5	0	0	$\frac{7}{5}$	0	$-\frac{1}{5}$	1	105	75
Tableau at end of 2nd and final iteration									
1c	Z	1	0	0	0	$+\frac{15}{7}$	$+\frac{16}{7}$	4,335	
2c	x_3	0	0	0	1	$-\frac{3}{7}$	$-\frac{13}{7}$	45	
3c	x_1	0	1	0	0	$\frac{2}{7}$	$-\frac{3}{7}$	270	
4c	x_2	0	0	1	0	$-\frac{1}{7}$	$\frac{5}{7}$	75	

3. Determine the new leaving basic variable by selecting the basic variable that becomes zero first as the entering basic variable is increased. To compute the upper bound for the entering basic variable for each constraint, divide the constant on the right-hand side of the equation by the coefficient of the entering basic variable in that equation. The equation that provides the smallest positive upper bound on the entering basic variable determines which basic variable leaves the basic.

4. Determine the new basic feasible solution by algebraically manipulating the equations such that the new basic variables and Z are expressed in terms of the new nonbasic variables.

5. Determine whether this solution is optimal by checking the coefficients of the nonbasic variables in the new Z equation. If all these coefficients are nonnegative, then this solution is optimal and the procedure stops. If there is a nonbasic variable with a negative coefficient, go to step 2.

Simplex Tableau Example for Maximization Problem (see Table 5.4)

1. The objective function has been rewritten as $Z - 13x_1 - 11x_2 = 0$; slack variables x_3, x_4, and x_5 have been added to the product mix problem, and the coefficients and constants are arranged in the initial tableau. The basic feasible solution is

$$\mathbf{X} = \begin{bmatrix} 0 \\ 0 \\ 1{,}500 \\ 1{,}575 \\ 420 \end{bmatrix}$$

with $Z = 0$.

2. The new entering basic variable, x_1, is identified by the smallest coefficient of a nonbasic variable from row 1a.

3. The upper bound of x_1 for each constraint is computed and shown in the last column of the initial tableau. The upper bound for row 2a is $1{,}500/4 = 375$; for row 3a, it is $1{,}575/5 = 315$; and for row 4a, it is $420/1 = 420$. Since 315 is the smallest positive upper bound, row 3a is the limiting constraint and x_4 is the leaving basic variable.

4. The simplex tableau at the end of the first iteration is obtained by:

a. Multiplying row 3a by $+13/5$ and adding the result to row 1a to obtain row 1b;

b. Multiplying row 3a by $-4/5$ and adding the result to row 2a to obtain row 2b;

c. Multiplying row 3a by 1/5 to obtain row 3b; and

d. Multiplying row 3a by $-1/5$ and adding the result to row 4a to obtain row 4b.

The new basic feasible solution is:

$$\mathbf{X} = \begin{bmatrix} 315 \\ 0 \\ 240 \\ 0 \\ 105 \end{bmatrix}$$

with $Z = 4,095$.

5. This solution is not optimal, because row 1b contains a negative coefficient for a nonbasic variable. Therefore, another iteration beginning with step 2 is necessary. It is left as an exercise for the student to show that the results of the next iteration produce the final tableau shown with optimal solution of

$$\mathbf{X} = \begin{bmatrix} 270 \\ 75 \\ 45 \\ 0 \\ 0 \end{bmatrix}$$

and $Z = 4,335$.

5.6 COMPLICATIONS IN APPLYING THE SIMPLEX METHOD

The solution procedure of the simplex method is based on the standard form of the linear programming model presented in earlier paragraphs. The simplex method will require modifications for use with forms other than the standard form. However, the concept of the simplex method still remains. Complications that may be encountered include:

1. An objective function to be minimized instead of maximized;

2. Inequality with "greater than," instead of "less than" sign;

3. Nonpositive constants on the right-hand side of the constraints;

4. Equalities instead of inequalities for constraints;

5. The requirements that a decision variable may be negative;

6. Tie between the variables when determining which variable will be the leaving basic variable;

7. More than one optimal solution; i.e., multiple solutions such that there is no unique optimal solution;

8. The constraints are such that no feasible solution exists;

9. The constraints are such that one or more of the variables can increase without limit and never violate a constraint; i.e., the solution is unbounded;

10. The requirement that the decision variables can only have integer values; and

11. Some or all of the coefficients and right-hand-side terms are given by a probability distribution rather than a single value.

The modifications necessary for overcoming complications 1, 2, and 4 are quite simple and readily accomplished and are discussed in the next section. Detailed treatment of the modifications necessary in each of the other cases can be found in most texts on linear programming, such as Dantzig (1963), Gass (1969), and Greenberg (1971) and in books on operations research, such as Wagner (1969).

5.7 DUALITY

After reviewing the results of the optimization study, the N. Dustrious Company now wants to know if it would be profitable to buy more raw material or work overtime so that more of its storage space could be used. The company currently pays $2 per pound for raw material and $240 per hour for labor at the plant.

One way to determine if an increase in activity is profitable is to compare the market value of a unit of resource with its shadow price. The shadow price is defined as the amount that the company could pay for one more unit of resource and just break even on the use of that unit in its operation. The shadow price, therefore, represents the value, to the company, of one more unit of the resource that could be used to produce output. If the shadow price of the resource is less than its market value, it would be unprofitable to purchase more of the resource because the company would lose money. However, if the shadow price of the resource is greater than its market value, it would be profitable to purchase and use more of the resource.

Let y_i represent the shadow price of the ith resource. Using shadow prices, the total value, to the company, of all resources used, P, is $1,500y_1$ for storage space plus $1,575y_2$ for raw material plus $420y_3$ for labor. The company will want to minimize P. Each unit of Product I requires 4 units of storage space, 5 units of raw material, and 1 unit of labor (from Table

5.1). Therefore, the value, to the company, of resources used per unit of product I is $4y_1$ for storage plus $5y_2$ for raw material plus y_3 for labor. The value, to the company, of the resources used per unit of output should not be less than the unit profit on the output, or some resource would be valued too low. This can be modeled by $4y_1 + 5y_2 + y_3 \geq 13$ where 13 is the unit profit for product I. Similarly, for product II, $5y_1 + 3y_2 + 2y_3 \geq 11$. Furthermore, the unit value of any resource cannot be negative. These conditions can be summarized in the following model:

Minimize
$$P = 1{,}500y_1 + 1{,}575y_2 + 420y_3$$
subject to:
$$4y_1 + 5y_2 + y_3 \geq 13$$
$$5y_1 + 3y_2 + 2y_3 \geq 11 \qquad (5.19)$$
$$\text{all } y_i \geq 0$$

The Dual Problem

Equations 5.19 appear to be a transformation of Equations 5.11; its constraint coefficients, for example, are the transpose of the constraint coefficients of Equations 5.11. If Equations 5.11 are called the primal problem, then Equations 5.19 are called the dual problem. The dual can be obtained directly from the primal as indicated by the following:

Primal Problem

Maximize

$$Z = c_1x_1 + c_2x_2$$

subject to:

$$k_{11}x_1 + k_{12}x_2 \leq b_1$$
$$k_{21}x_1 + k_{22}x_2 \leq b_2$$
$$k_{31}x_1 + k_{32}x_2 \leq b_3$$
$$\text{all } x_i \geq 0$$

Dual Problem

Minimize

$$P = b_1y_1 + b_2y_2 + b_3y_3$$

subject to:

$$k_{11}y_1 + k_{21}y_2 + k_{31}y_3 \geq c_1$$
$$k_{12}y_2 + k_{22}y_2 + k_{32}y_3 \geq c_2$$
$$\text{all } y_i \geq 0$$

Every linear programming model has a dual associated with it. The dual of the dual problem is the primal. Hence, the interpretation of the dual varies from one application to another.

Solving Minimization Problems

In order to apply the simplex method to minimization problems, the way in which the initial basis is selected and the criterion for a new incoming basic variable must be modified.

When the linear programming model is the standard form, as shown in Equations 5.3, with all $b_i \geq 0$ slack variables are added to the left-hand side of the constraints to form equalities. The slack variables are then used as the initial basic variables. However, consider the inequalities in Equations 5.19. The slack variables, which must be positive, must be subtracted from the left-hand side of the inequalities in order to transform them into equalities; i.e.,

$$4y_1 + 5y_2 + y_3 - y_4 = 13$$
$$5y_1 + 3y_2 + 2y_3 - y_5 = 11$$

$$(5.20)$$

An initial basis with $y_4 = -13$ and $y_5 = -11$ is not feasible and cannot be used because the simplex method must start with a feasible solution. The following is one way to obtain an initial basis once the constraints have been transformed into equalities.

Add another variable with a coefficient of $+1$ to the left-hand side of the equation. Hence, Equations 5.20 become

$$4y_1 + 5y_2 + y_3 - y_4 + y_6 = 13$$
$$5y_1 + 3y_2 + 2y_3 - y_5 + y_7 = 11$$

$$(5.21)$$

where y_6 and y_7 are called artificial variables. The name artificial variables is given to y_6 and y_7 because they are added as an artifice in order to obtain an initial basic solution.

A feasible solution with $y_6 = 13$ and $y_7 = 11$ can now be obtained. However, this solution has no meaning in the original problem since the artificial variable is merely an artifice with no meaning. Therefore, an artificial variable must be assigned a penalty that will insure that it will not be in the optimal solution. This can be accomplished for a minimization problem by assigning the artificial variable a sufficiently large positive coefficient in the objective equation. The objective function in Equations 5.19 then become

Minimize

$$P = 1{,}500y_1 + 1{,}575y_2 + 420y_3 + 5{,}000y_6 + 5{,}000y_7 \qquad (5.22)$$

For maximization problems, the artificial variable would be assigned a large negative coefficient in the objective function.

The coefficients and constants in Equations 5.21 and 5.22 can now be arranged in the initial tableau, as shown in Table 5.5.

Table 5.5

SUMMARY OF SIMPLEX TABLEAU PROCEDURE FOR MINIMIZATION PROBLEM

ROW NO.	BASIC VARIABLE	P	y_1	y_2	y_3	y_4	y_5	y_6	y_7	RHS	UPPER BOUND ON ENTERING VARIABLE
Initial tableau											
1a	P	1	$-1{,}500$	$-1{,}575$	-420	0	0	$-5{,}000$	$-5{,}000$	0	
2a	y_6	0	4	5	1	-1	0	1	0	13	13
3a	y_7	0	5	3	2	0	-1	0	1	11	11
Second tableau											
1b	P	1	$43{,}500$	$38{,}425$	$14{,}580$	$-5{,}000$	$-5{,}000$	0	0	$120{,}000$	
2b	y_6	0	4	5	1	-1	0	1	0	13	$\frac{13}{4}$
3b	y_7	0	5	3	2	0	-1	0	1	11	$\frac{11}{5}$
Third tableau											
1c	P	1	0	$12{,}325$	$-2{,}820$	$-5{,}000$	$3{,}700$	0	$-8{,}700$	$24{,}300$	
2c	y_6	0	0	$\frac{13}{5}$	$-\frac{3}{5}$	-1	$\frac{4}{5}$	1	$-\frac{4}{5}$	$\frac{21}{5}$	$\frac{21}{13}$
3c	y_1	0	1	$\frac{3}{5}$	$\frac{2}{5}$	0	$-\frac{1}{5}$	0	$\frac{1}{5}$	$\frac{11}{5}$	$\frac{11}{3}$
Fourth tableau											
1d	P	1	0	0	24.23	-259.62	-92.31	$-4{,}740.38$	$-4{,}907.69$	$4{,}390.38$	
2d	y_2	0	0	1	$-\frac{3}{13}$	$-\frac{5}{13}$	$\frac{4}{13}$	$\frac{5}{13}$	$-\frac{4}{13}$	$\frac{21}{13}$	$\frac{21}{13}$
3d	y_1	0	1	0	$\frac{7}{13}$	$\frac{3}{13}$	$-\frac{3}{13}$	$-\frac{3}{13}$	$\frac{5}{13}$	$\frac{11}{13}$	$\frac{16}{7}$
Final tableau											
1e	P	1	-45	0	0	-270	-75	$-4{,}730$	$-4{,}925$	$4{,}335$	
2e	y_2	0	$\frac{3}{7}$	1	0	$-\frac{2}{7}$	$\frac{1}{7}$	$\frac{2}{7}$	$-\frac{1}{7}$	$\frac{15}{7}$	
3e	y_3	0	$\frac{13}{7}$	0	1	$\frac{3}{7}$	$-\frac{3}{7}$	$-\frac{3}{7}$	$\frac{5}{7}$	$\frac{16}{7}$	

The coefficients of the basic variables must be zero in the objective function row (row 1) before a check for optimization can be made. Therefore, algebraically manipulate the equations such that the basic variables y_6 and y_7, and P are expressed in terms of the nonbasic variables as shown in the second tableau in Table 5.5.

Instead of selecting the entering variable to increase the value of the objective function as in the maximization problem, the new entering variable is selected as the variable that causes the greatest rate of decrease in the objective function. This is accomplished by selecting the variable in row 1 with the largest positive value as the entering variable.

The modification in the criterion for selecting the entering variable leads to the result that the optimal solution has been obtained when all nonbasic variables have nonpositive coefficients in the objective function row. The other steps in the simplex method remain unchanged. The solution procedure for the minimization problem is summarized in Table 5.5.

Interpreting Results from Dual

The optimal solution of $y_1 = 0$, $y_2 = \frac{15}{7}$ and $y_3 = \frac{16}{7}$ indicate that the shadow price for storage space is zero, for raw materials is $15/7 per pound and for labor is $16/7 per minute. The labor is in units of minutes because of the way in which Equation 5.4 was derived.

The shadow price of zero for storage space means that the value of additional storage space to the company is zero. This is reasonable because it already has more space than it can use.

However, the company could afford to pay $15/7 for a pound of raw material and still break even. Since $15/7 is greater than the market value of $2 per pound, it is profitable to buy more material and increase production. The increase in profit to the company will be $(15/7 - 2)$ or $1/7 per pound of additional material used. The company can continue to purchase more material and increase production and profits until one of the other constraints becomes effective. The amount of additional material that can be used before another constraint becomes effective can be determined by extending the simplex method. The method is explained in most textbooks on linear programming such as Dantzig (1963) and Gass (1969).

The shadow price for labor of $16/7 per minute is less than the $4 per minute that the company must pay to keep the plant running; therefore, it is not profitable to work overtime.

Comparing the last tableau in Table 5.5 with the last tableau in Table 5.4 reveals that the value of the basic variables in the dual solution is equal to the coefficients of the nonbasic variables in row 1 of the primal solution. Also, the optimal values of the basic variables in the primal are the coefficients of the nonbasic variables in row 1 in the dual solution with their signs changed, and the optimal value of Z is equal to the optimal value of P. Hence, it is not necessary to solve both the primal and the dual since the solution to one contains the solution to the other. Which method to use is a matter of convenience or computational efficiency since the dual and the primal problems are not always equally easy to solve. Usually, an additional constraint requires more computational effort than an additional variable. Thus, if the primal should have a large number of constraints and relatively few variables, its dual will probably require less computation since the number of variables and constraints are interchanged.

5.8 NONLINEAR PROGRAMMING

Nonlinear programming models are similar to linear programming models except that the objective function and constraining equations are not required to be linear functions of the decision variables.

The Product Mix Problem Again

In the product mix problem of Section 5.3, which was formulated as a linear programming model as shown in Equation 5.4, it was assumed that the average profit per unit of product was constant regardless of how many units of a product were produced. Sometimes, however, the profit per unit depends on the number of units produced.

Assume that the average profit per unit of product I decreases as each additional unit is produced and that the profit per unit is given by $\$(40. - 0.1x_1)$, where x_1 is the total number of units of product I. If all other problem information remains the same, the total profit for the company is now

$$Z = (40. - 0.1x_1)x_1 + 11x_2$$

or

$$Z = 40x_1 - 0.1x_1^2 + 11x_2 \tag{5.23}$$

This profit must be maximized subject to the constraints of Equation 5.4. Thus, the problem is formulated as a nonlinear programming problem that has a nonlinear objective function and linear constraints.

Other types of nonlinear models that can be formulated include models with a linear objective function and one or more nonlinear constraints and models with a nonlinear objective function and one or more nonlinear constraints.

Solving Nonlinear Problems

Many nonlinear problems can be solved by techniques that are slight variations of those applied to the linear model. However, some problems may require much more sophisticated techniques. At present, there exists no efficient solution procedure for the general nonlinear problem. Substantial progress has been made, however, for some special cases.

A problem with only a few variables can be solved graphically. The graphical solution process for nonlinear problems is analogous to that for the linear problem. The graphical solution for the nonlinear product mix problem formulated above is summarized in Figure 5.4. The only difference between this solution procedure and that for the linear problem is that the line representing profit is now nonlinear instead of linear. The maximum profit is \$5,285.60, with $x_1 = 172$ and $x_2 = 124$.

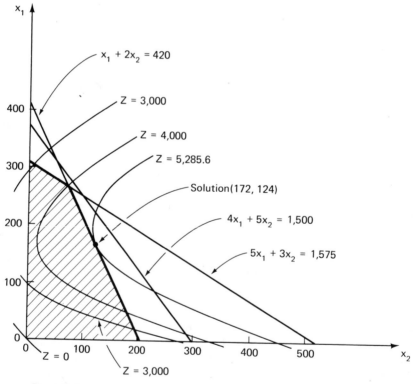

Figure 5.4

Graphical Solution for Nonlinear
Programming Problem

There is no single procedure for solving all nonlinear programming problems and it is difficult to provide a systematic listing of available algorithms. The success of a particular algorithm depends on the number of variables and constraints and the type of computation facility available. All procedures are iterative and many involve calculating first and, in some cases, second derivatives of the objective function.

One of the most useful nonlinear programming techniques is quasi-linearization. The application of this technique requires that the non-linear objective function be approximated by linear segments and that the constraints be linear. The advantage of this method is that it uses a modification of the simplex method, and computer codes are available for its use (IBM, 1969).

A detailed discussion of the techniques of nonlinear programming is beyond the scope of this book; however, a unified approach to nonlinear programming has been presented by Wilde and Beightler (1967).

5.9 SUMMARY

Linear programming is a technique for optimizing the allocation of resources subject to constraints. The linear programming models developed in this chapter and in Section 3.5 illustrate the variety of problems for which it is suited.

The development of the criterion function and the constraint equations constitutes the professional aspect of problem-solving. Once the model has been formulated there are algorithms, such as the simplex method, that are available for solving the model.

Each linear programming problem is inherently associated with the formulation of a dual problem. The value of the optimal solution for a primal problem is identical to the optimal solution for its dual. The information from the dual problem can be most helpful to the analyst because of its implications for sensitivity analysis. The variables in the dual are the shadow prices of the constraints in the primal. The value of the shadow prices at the optimum indicate the value of additional units of the scarce resources, and thus, the effect of the constraints on the optimal solution.

Therefore, the solution tableau for the simplex method indicates the optimal values for the decision variables, the constraints that are effective, and the value of additional units of the limited resources. This last value indicates how sensitive the optimal solution is to the quantity of the limited resource.

There is no solution algorithm available for nonlinear programming which has the applicability that the simplex method has in linear programming. However, there are specialized solution techniques for certain kinds of nonlinear problems, and those problems with only a few variables can be solved graphically.

5.10 PROBLEMS

P5.1. Listed below are the objective function and constraints. Determine the maximum value of Z subject to the constraints. Work this problem graphically and by the simplex method.

1. $2x + 3y \leq 48$
2. $3x + 2y \leq 42$
3. $5x + y \leq 50$
4. $x - 3y \leq 0$
5. $x \geq 0; y \geq 0$

$$Z = 6x + 2y$$

P5.2. A contractor is planning a job that will require a large amount of gravel and sand. He estimates that he will need:

coarse gravel	20,000 cu yd
fine gravel	29,000 cu yd
sand	20,000 cu yd

There are two pits from which he can obtain material and he plans to haul from these two pits and separate (screen) on the job the material he needs. Analysis shows that the material at each pit has the following composition:

	PIT A	PIT B
coarse gravel	20%	30%
fine gravel	14%	50%
sand	25%	20%
waste	41%	0%

It costs him \$8/cu yd for material and hauling from Pit A and \$16/cu yd from Pit B. How much material should be hauled from each pit to minimize the hauling cost? What will the total hauling cost be? Which constraint is slack? What does this slack constraint mean in terms of material on the job? Solve this problem graphically and by the simplex method.

P5.3. Develop mathematical programming models for Problem 2.3.

P5.4. A manufacturing plant requires 300,000 gallons of water per day. The water is filtered before it is delivered to the plant. The manufacturing process requires that the water be chlorinated and softened before it can be used. Two different brands of water additive contain some amount of a chlorination chemical, SMILO, and a softening agent, LATHERO. One package of the product put out by the Hooker Drug Co. contains eight lb of LATHERO and three lb of SMILO. One package of the Amerikan Kemical Korp. product contains four lb of LATHERO and nine lb of SMILO.

One-hundred and fifty lb. of LATHERO and 100 lb of SMILO are needed daily to maintain the water at an acceptable level of softness and to meet recommended levels of chlorination. A package of the Hooker Drug Co. product costs $8 and a package of the American Kemical Korp. product costs $10. What should be the plant manager's policy to meet the requirements for the water additives at least cost?

P5.5. Assume that the type of occupancy for a high-rise building is to be decided and you are asked to determine the number of units of each type that will maximize the profit. You are to provide the objective function and the constraints for a linear programming problem. The occupancy that you are to consider is the following:

1. First floor will be used for entrance, coffee shop, etc.,

2. One floor for offices for the company (second floor),

3. Hotel rooms,

4. One-room efficiency apartments,

5. Two-bedroom apartments,

6. General office space,

7. Restaurant,

8. Basement (recreational space for occupants plus a meeting room).

How would you approach the problem, what kind of information would you require, and how would you establish the objective function and constraints?

P5.6. A reservoir is designed to provide hydroelectric power and water for irrigation. The outflow or release from the reservoir is used to generate hydroelectric power and after passing through the turbines it can be diverted toward an irrigation project or it can continue to flow downstream. At least 1 unit of water must be left to flow down the stream each month. The reservoir has a capacity of 10 units of water. It currently contains 5 units of water and must contain 5 units of water at the end of the year. The maximum amount that can be released in any month is 7 units. The return for each unit of water released for hydroelectric power and irrigation and the estimated inflow for each month of a year are shown in Table P5.6.

Table P5.6

	JAN.	FEB.	MAR.	APR.	MAY	JUNE	JULY	AUG.	SEPT.	OCT.	NOV.	DEC.
Inflow, units of water	2	.2	3	4	3	2	2	1	2	3	3	2
Hydroelectric power 10^6/unit of water released	1.6	1.7	1.8	1.9	2.0	2.0	2.0	1.9	1.8	1.7	1.6	1.5
Irrigation 10^6/unit of water released	1.0	1.2	1.8	2.0	2.2	2.2	2.5	2.2	1.8	1.4	1.1	1.0

Formulate the linear programming model that would be solved to determine how much water to release each month of the year so that the returns from hydroelectric power and irrigation will be maximized.

P5.7. A concrete transit-mix company owns three plants with capacities and production costs as follows:

PLANT NUMBER	DAILY CAPACITY (YDS)	PROD. COST ($/YD)
I	160	10
II	160	9
III	80	13

The company is under contract to supply concrete for a bridge construction and is scheduled to deliver to the various job sites the following quantities of concrete:

JOB SITE	AMOUNT (YDS)
1. south bank pier	100
2. south bank abutment	40
3. north bank abutment	80
4. north bank pier	140

Based on distance, traffic, and site delays, the following haul costs are estimated.

HAUL COST, ($/YD)

Job Site Number	Plant Number		
	I	II	III
1	2	1	3
2	1	2	3
3	2	1	2
4	3	2	1

Schedule tomorrow's production to minimize total cost to the company.

P5.8. Give reasons why you agree or disagree with the following:

1. In mathematical programming problem formulation, the decision sequence in model building is to first define the objective function, then the decision variables, and finally find the constraints.

2. In general, the number of constraints in a mathematical programming problem formulation is unknown.

3. In Section 5.4 the problem solution is not affected by the addition of the new constraint $x_1 + 0.5x_2 \geq 100$.

4. In Section 5.4 the problem solution is not affected by the addition of the new constraint $x_1 + 0.5x_2 \leq 100$.

5. None of the preceding problems can be solved more efficiently using the dual formulation.

6

ORGANIZATIONAL NETWORKS

Glen Canyon Bridge, Dam and Powerplant in northern Arizona was selected as the Outstanding Civil Engineering Achievement of 1964. The powerplant is designed to generate electricity for 1.5 million people in areas of Arizona, Colorado, New Mexico, Utah, and Wyoming. Over five million cubic yards of concrete were used in the dam's construction. At the peak of construction, 2,300 men were employed at the site. The city of Page, Arizona was constructed to house those who built and maintain the project. (Courtesy of the American Society of Civil Engineers)

6.1 AN ORGANIZATIONAL SYSTEMS CONCEPT

Engineers are often responsible for planning and supervising projects. Although most projects are associated with the construction of engineering, industrial, and housing works, others are related to the implementation of commercial products, advertising campaigns, research programs, planning studies, and the like. In all cases the projects can be defined as a collection of work tasks or activities, each of which must be performed before the project can be completed.

Each work task entails selecting, evaluating, and enumerating the resources (material, equipment, labor, finance, etc.) necessary to insure its completion. In addition, it is necessary to determine for each work task the conditions that must exist for the task to begin. Finally, procuring the necessary resources and using them at a specific time requires a major scheduling effort for the project.

Planning a project involves selecting the technological methods to be used and determining a specific work order from all the various ways and sequences in which the job could be done.

Scheduling a project requires determining the timing of the work task activities that comprise the project and coordinating them so that the overall project time can be determined.

Supervising a project involves conceiving, implementing, and monitoring an information system that will permit the project status to be evaluated at any time. Such an information system model, coupled with a planning model, can be the base on which project control can be established or on which corrective action can be initiated, if necessary, to insure the smooth progression of the project as planned.

In all these organizational phases the engineer is concerned with identifying components (work tasks) and structuring them into a coherent whole system. The obvious organizational aspects of these efforts suggest that the system as produced be called an organizational system.

A modeling method for organizational systems associated with project planning, scheduling, and supervision was introduced to the construction industry about 1960 and is now commonly known as the *Critical Path Method* (CPM). Critical path methods require that linear graph models be formulated to specifically represent the unique features and *plan for the project* under consideration. The various project activities are collected together and synthesized in a connected linear graph to show the project logic and the sequential nature of the various activities. Once the project plan is defined through the CPM structural graph model, a variety of attributes associated with the project work tasks can be added to the model. In this way organizational problems can be modeled as simple graphical problems.

A simple organizational problem is that concerned with the sequencing of work tasks (or activities) when each task has physical or technological prerequisites that must be met before the task can begin (i.e., scheduled). Thus, sequencing activities leads to directed path network models.

6.2 THE CRITICAL PATH METHOD

The critical path method is a graphical process requiring the development of a graph model for the project. The project work task activities become the system components, and the system structure is determined by the technological and managerial ordering and sequencing of the project activities.

There are two ways of modeling a relation between two elements using linear graphs. If the elements are modeled as nodes, the relation between them can be portrayed as a directed edge. This concept is the basis for the circle notation (activity on node) representation of projects, and is commonly used in CPM computer programs. If the elements are modeled as branches with the common node between them indicating the relation between the elements, then the familiar and popular arrow diagram (activity on branch) representation of construction project results.

As an example, the following problem is concerned with the construction of a garage and a driveway. The garage is to be of wood frame construction with a concrete floor, and the driveway is to be of concrete.

A first step in the CPM analysis is to compile a list of activities that collectively comprise the project. The number and extent of the activities defined depends on the project, the level of detail required, and the intended management use of its model. For this project a total of fourteen activities have been indicated as being necessary to complete the project. In addition, the time required to complete each activity has been estimated. The activities and the estimated times are given in Table 6.1.

It is possible to further subdivide each activity if a more detailed analysis is desired. For example, constructing the concrete slab for the garage could be broken into forming, acquiring concrete, pouring slab, and finishing slab. The same is true for the other activities. The degree of breakdown of activities depends on the analytic detail that is required.

Given the list of activities, the logic associated with their accomplishment must be determined. In this case clearing and preparing the site and obtaining material could occur simultaneously and anytime after

Table 6.1

LIST OF ACTIVITIES AND ESTIMATED TIMES FOR GARAGE AND DRIVEWAY PROJECT

ACTIVITY	REQUIRED TIME (WORKING HOURS)
1. Clear and prepare site	10
2. Obtain materials	8
3. Construct concrete slab for garage	6
4. Prefabricate wall panels and roof trusses	16
5. Cure concrete slab for garage	24
6. Erect walls	4
7. Erect roof trusses	4
8. Put on siding and windows	10
9. Install door	4
10. Put on roofing	12
11. Paint	16
12. Construct concrete slab for driveway	8
13. Cure concrete slab for driveway	24
14. Clean up site	4

the start of the project. Both would have to be completed before the concrete slab for the garage could be constructed. Prefabricating the wall panels and trusses could be done any time after the materials are obtained and is independent of the slab construction. The wall panels cannot be erected until the concrete slab has cured and must precede the erection of the roof trusses.

For this project, it was decided to separate construction of the garage floor slab from the driveway because the concrete trucks could not get to the garage site if the forms were in place for the driveway; thus the garage slab must be poured before the driveway activity can commence. However, construction of the garage may proceed independent of the driveway except for the final clean up activities.

After the garage wall panels are erected, the siding and windows can be put on and the door installed. These activities could occur simultaneously with the erection of the roof trusses and the installation of roofing. The painting, however, can be done only after these activities are completed. Finally, the site can be cleaned up after the driveway has cured and the garage has been completed.

This completes the definition of both the system components and structure as defined by the rationale of the construction plan and technology.

Linear graph models for the project are shown in Figure 6.1. Figure 6.1a is a circle notation model in which activities are modeled on nodes and the system structure of the construction plan is indicated by directed branches. Figure 6.1b is an arrow notation model in which activities are modeled by directed branches and the system structure is indicated by the nodes to which the branches are attached. The dashed arrow between nodes B and C is a logic branch to indicate that activity 2 must be completed before activity 3 can begin. This arrow is called a dummy activity and it has zero activity duration. Activities can also be referred to by the nodes to which they are connected, i.e., activity 2 can be called activity A-B or activity AB.

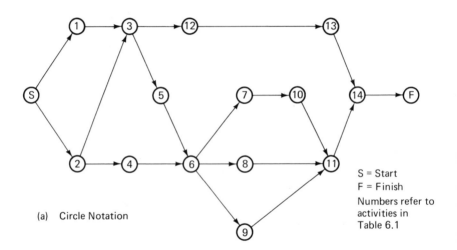

S = Start
F = Finish
Numbers refer to activities in Table 6.1

(a)   Circle Notation

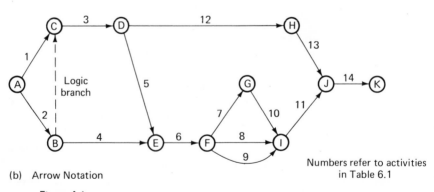

Numbers refer to activities in Table 6.1

(b)   Arrow Notation

Figure 6.1

Linear Graph Models of Garage Project

6.3 LINEAR GRAPH PROPERTIES AND ANALYSIS

Assigning times for completing each project activity enables the linear graph model to serve as a scheduling model. Thus it is possible to determine the following:

1. The earliest time at which an activity may be started (known as earliest start time or EST);

2. The minimum time in which the total project may be completed;

3. The latest time at which an activity may be started if the project is to be completed in a minimum time (known as latest start time or LST);

4. The earliest time at which an activity may be finished (known as the earliest finish time or EFT); and

5. The latest time at which an activity may be finished if the project is to be completed in a minimum time (known as the latest finish time or LFT).

Each of these pieces of information is important in scheduling an actual project, because they define when a particular activity may occur within the overall work schedule.

The earliest start time represents the earliest time that a given activity can begin after the initiation of a project. This time is a function of the other activities that must be completed prior to starting the particular activity under consideration. The earliest start time can be ascertained by determining the maximum necessary time required for preceding activities. This requires summing the time requirements along each linear graph path from the starting point to the activity involved.

The minimum time in which the project can be completed is given by the duration of the maximum earliest start time path. That earliest start time path from the start of the project which determines the project completion time is known as the critical path.

Determining the EST time requires the selective addition of activity durations along the various directed paths to each node of the network. The combinatorial calculation can be systematically developed so that only the set of preceding nodes need be considered at any time.

Let T^E represent the earliest start time at node i. For the arrow notation network of Figure 6.2b the earliest start time at node B is determined by the path "from node A" and is the earliest start time at node A plus the duration of the activity between nodes A and B (ac-

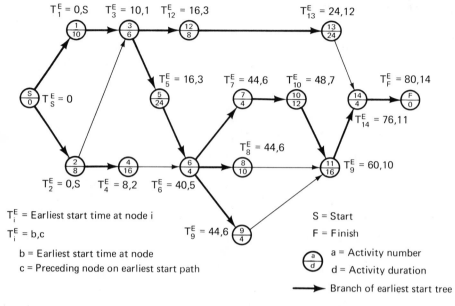

T_i^E = Earliest start time at node i

T_i^E = b,c

  b = Earliest start time at node

  c = Preceding node on earliest start path

S = Start

F = Finish

a = Activity number

d = Activity duration

→ Branch of earliest start tree

(a) Circle Notation

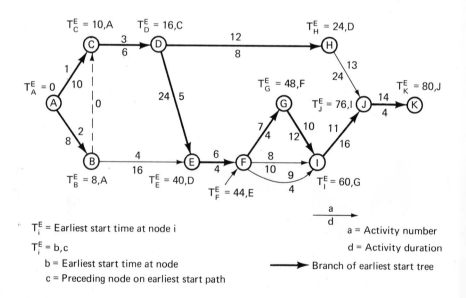

T_i^E = Earliest start time at node i

T_i^E = b,c

  b = Earliest start time at node

  c = Preceding node on earliest start path

a = Activity number

d = Activity duration

→ Branch of earliest start tree

(b) Arrow Notation

Figure 6.2

Earliest Start Trees of Garage Project

tivity 2). Therefore,

$$T_B^E = T_A^E + 8 = 8 \text{ from node } A$$

$$T_C^E = \text{maximum} \begin{cases} T_A^E + 10 = 10 \\ T_B^E + 0 = 8 \end{cases} \qquad \therefore T_C^E = 10 \text{ from node } A$$

$$T_D^E = T_C^E + 6 = 10 + 6 = 16$$

$$T_E^E = \text{maximum} \begin{cases} T_D^E + 24 = 40 \\ T_B^E + 16 = 24 \end{cases} \qquad \therefore T_E^E = 40 \text{ from node } D$$

(6.1)

Finally $T_K^E = 80$ indicates that the project cannot be completed in less than 80 working hours.

It is convenient to include in the tabulation the label for the unique preceding node on the directed path that determines the earliest start time for the specific node under consideration. The calculations are left to the reader and are illustrated in Figure 6.2 for both circle and arrow notation graphs. In this way the preceding calculations and labels identify the earliest start time paths for each node. The collection of these paths produces a special graph structural property called a directed tree. In Figure 6.2 the directed trees are shown in both graphs by the heavy branches. For obvious reasons the trees are called the earliest start trees of the CPM network.

In linear graph theory a tree of a connected graph is a minimal collection of branches such that the graph is still connected and only one path exists between any two nodes. In CPM graphs the branches are directed in such a way that the tree produced can be said to be rooted at the datum node.

The latest start time represents the latest time that an activity may be initiated in the schedule if the project is to be completed in a minimum time. Again, the LST is specified in terms of the time from the start of the project; however, it is computed backwards in time from the project finish node based on the minimum project time.

The actual LST is calculated by determining the minimum time path from the completion of the project. If T_j^L is the latest start time at node j, then from Figure 6.3 T_J^L is T_K^L minus the duration of activity 14 (i.e., $T_J^L = T_K^L - 4 = 80 - 4 = 76$). The latest start time at node K is the same as the earliest start time because node K represents the end of the project. Also, plotting minimum time paths for all nodes from the completion node of the project results in the latest start tree, as shown in Figure 6.3. Its derivation is left to the reader.

Figure 6.4 illustrates the calculations that might be made for each activity in the CPM network. The EFT for activity ij is computed as the sum of the EST(T_i^E) and the activity duration. The LST for activity ij is computed as the difference between LFT (T_j^L) and the activity

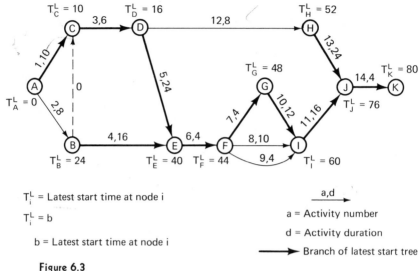

T_i^L = Latest start time at node i

T_i^L = b

   b = Latest start time at node i

a = Activity number

d = Activity duration

⟶ Branch of latest start tree

Figure 6.3

Latest Start Tree for Arrow Notation Model of Garage Project

duration. All activities that start at the same node have the earliest start time of that node. Also all activities that end at the same node have the latest finish time of that node.

The activities that take place on the critical path for the project will have an early start time that will be the same as the latest start time.

The critical path that determines the minimum project duration passes through activities 1, 3, 5, 6, 7, 10, 11, and 14. The summed duration of these critical activities equals 80 working hours. The actual duration of the project will depend on when the activities are scheduled and worked and whether consecutive days are used.

For activities not on the critical path, however, the earliest start time will be less than the latest start time. The difference between the two is known as the float time. A variety of measures of float time have been developed and used. Thus if an arrow notation activity A_{IJ} of duration d_{IJ} is considered joining nodes I and J and T^E and T^L are earliest and latest start node time symbols, then the following float definitions can be made:

$$\left\{\begin{array}{l} \text{Total float } TF_{IJ} = T_J^L - (T_I^E + d_{IJ}) \\ \text{Free float } FF_{IJ} = T_J^E - (T_I^E + d_{IJ}) \\ \text{Interfering float} = TF_{IJ} - FF_{IJ} \\ \qquad\qquad\qquad = T_J^L - T_J^E \end{array}\right. \qquad (6.2)$$

Total float for an activity corresponds to the concept of making available to the activity the greatest amount of available float time without jeopardizing the project duration. It, therefore, assumes T^L at node J and T^E at node I. It can be used up only once on the chain passing through the activity.

Free float for an activity corresponds to the concept of making available to the activity only that amount of available time that does not interfere with subsequent activities. It, therefore, assumes T^E times at both nodes I and J.

The use of interfering float by, and during, an activity indicates that subsequent activities are affected in that they can no longer utilize all their previously available total float.

Figure 6.4 also shows the various floats for the garage project. Figure 6.5 shows the relationships between the different float measures.

A time-scaled version of the CPM model is commonly used in construction practice and is called a Bar, or Gantt, Chart. The bar chart for the garage project is shown in Figure 6.6, drawn to a time scale of work hours assuming all activities are commenced as soon as possible. Of course the activities must be treated in calendar time, using normal work days of eight-hour units, unless overtime and/or weekend construction is contemplated.

The bar chart of Figure 6.6 reveals that during hour 46, for example, activities 7, 8, 9, and 13 are being undertaken simultaneously if the earliest start schedule is being followed. The various resources utilized by these activities are therefore simultaneously required at that time. This observation enables a cut-set concept to be applied to organizational networks in which capacity flows correspond to the resource use concepts.

6.4 RESOURCE SCHEDULING

A further dimension to the model is thus added to the modeling of a project by considering resources. These resources might be labor, cash, equipment, or materials. Limited resources actually are additional constraints on the conduct of the project. The constraints may be due to one of the following reasons:

1. There may only be a limited amount of a particular resource available. The amount of equipment or the number of men with certain skills are usually limited. This limitation may be due to a shortage of supply or it may be imposed by the financial resources of the contractor.

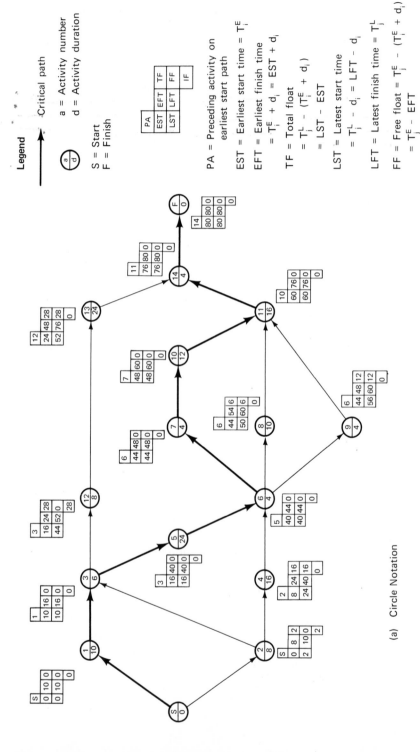

(a) Circle Notation

PA = Preceding activity on earliest start path

EST = Earliest start time = T_i^E

EFT = Earliest finish time = $T_i^E + d_{ij}$

TF = Total float = $T_j^L - (T_i^E + d_{ij}) = LFT - (EST + d_{ij})$
     = LST − EST

LST = Latest start time = $T_j^L - d_{ij}$

LFT = Latest finish time = T_j^L

FF = Free float = $T_j^E - (T_i^E + d_{ij}) = T_j^E - EFT$

IF = Interfering float = TF − FF

a = Activity number, the activity can
    also be designated by the nodes
    it connects
d = Activity duration

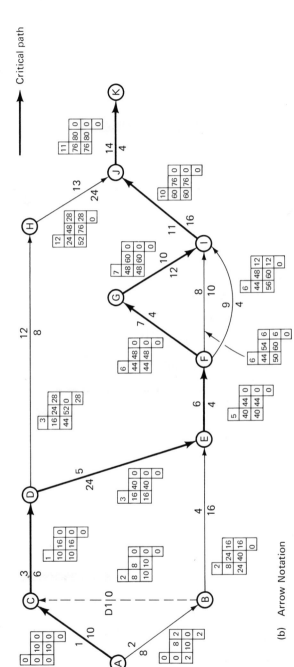

(b)   Arrow Notation

Network Calculations for Garage Project

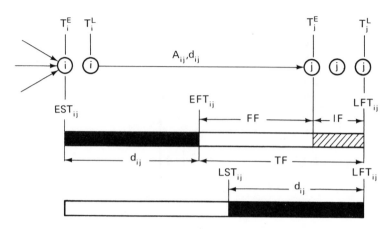

Figure 6.5

Relationships Between Floats

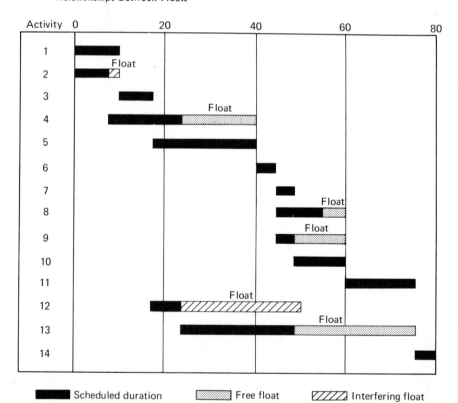

Figure 6.6

Bar Chart for Garage Project for Earliest Start Times

2. It may be desirable to keep the level of resource usage nearly constant for the duration of its use on the project. To avoid frequent hiring and firing of men it is desirable to keep the level of manpower stable for the duration of the project. Also, it may be desirable to schedule the use of certain equipment, such as a crane or concrete mixing plant, so that its use will be continuous and nearly constant for the time that it must be at a project site.

Consider the problem of scheduling the activities involved in constructing a reservoir that will be used to supply water and recreation facilities and to generate electricity.

The list of activities and estimated durations given in Table 6.2

Table 6.2

LIST OF ACTIVITIES AND ESTIMATED DURATION AND NUMBER OF MEN REQUIRED FOR EACH ACTIVITY IN RESERVOIR CONSTRUCTION

ACTIVITY NUMBER	ACTIVITY	DURATION (MONTHS)	MEN REQUIRED
1	Clear dam site	2	10
2	Clear rest of reservoir area	34	25
3	Construct earth dam	40	60
4	Construct concrete spillway	20	30
5	Powerhouse structure	18	20
6	Install generators	10	10
7	Construct access roads	3	20
8	Construct other roads	8	30
9	Construct recreation facilities	34	35
10	Administration building and utilities	5	10
11	Water treatment	10	15
12	Install plant equipment	12	5

are used for the preliminary planning of the project. The CPM arrow network is presented in Figure 6.7 along with the durations of the activities. The critical path is composed of activities 7, 1, 3, 5, and 6 and the minimum project duration is 73 months. The critical path bar chart based on earliest start times is shown in Figure 6.8.

Assume that the number of men required to complete each activity in Figure 6.7 is that shown in Table 6.2. Furthermore, assume that the men can work on any activity; i.e., they are not specialists. The total

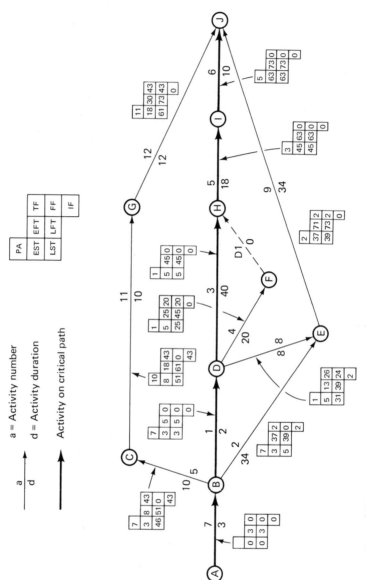

Figure 6.7

Arrow CPM Network for Reservoir Construction

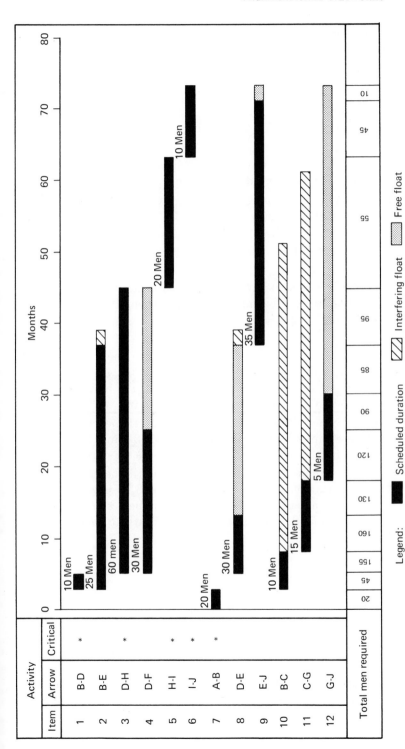

Figure 6.8

Bar Chart for Reservoir Construction for Earliest Start Times

number of men required during each month can be determined by adding the number of men required for each activity that is active that month. The values are shown at the bottom of Figure 6.8.

If there are 160 men available to work on the project, the early start schedule is satisfactory. However, if only 120 men are available, then a new schedule must be computed.

One heuristic method for determining the new schedule is to:

1. Start with the first day and schedule all jobs possible, then do the same for the second day, and so on.

2. If several jobs compete for the men, then schedule first the one with the smallest float.

3. Then, if possible, reschedule activities not on the critical path in order to free the men for the critical path activities.

Figure 6.9 is a revised bar chart that meets the requirement that not more than 120 men can be employed at any time. It was developed according to the following steps:

1. One job can be started the first month, activity 7. It is scheduled and requires 20 men. The bar for the activity is shaded for the duration of the activity and the number of men is indicated at the upper-left corner of the bar.

2. No other activities can begin until the fourth month; activity 7 has been completed and activities 1, 2, and 10 can start. Because they require a total of only 45 men, they are all started.

3. The next new activity cannot begin until the sixth month. Activity 1 has been completed and activities 3, 4, and 8 can begin. If 3, 4, and 8 are all started and 2 and 10 remain active, 155 men would be required. Comparing the total floats for each activity:

Activity	Men	Float	
3	60	0	← smallest float
4	30	20	
8	30	26	

Therefore, schedule activity 3 with a total of 95 men employed.

4. At the beginning of month 9, activity 10 is completed and

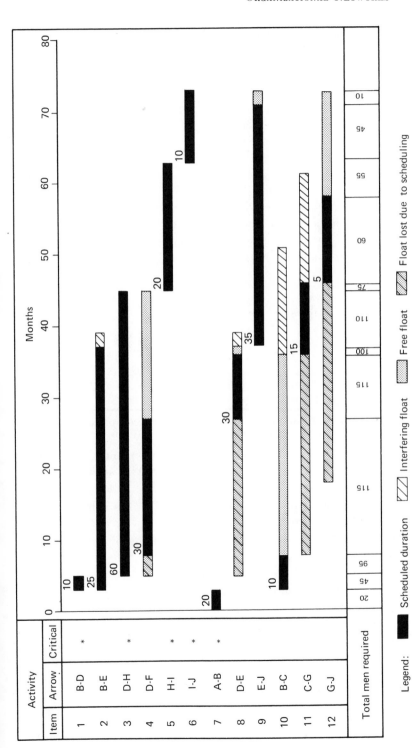

Figure 6.9

Bar Chart for Labor Constrained Schedule for Reservoir Construction

this makes 10 more men available. Activities 4 or 8 or 11 can start but not all three. The float comparison is:

Activity	Men	Float	
4	30	17	← smallest float
8	30	23	
11	15	43	

Therefore, schedule activity 4. Note that activities 4 and 8 have lost 3 months of float at the beginning of month 9.

5. Activity 4 is completed at the start of month 29, making 30 more men available. Activities 8 and 11 can be started.

Activity	Men	Float	
8	30	3	← smallest float
11	15	23	

Therefore, schedule activity 8.

6. At the start of month 37, activity 8 is completed, releasing 35 men. Activity 11 is started.

7. Activity 2 is completed at the start of month 38 and this makes 25 more men available. Activity 9 is started, leaving 10 men available.

8. At the beginning of month 46, activity 3 is completed and activity 5 is started.

9. Activity 11 is completed at the beginning of month 47 and activity 12 is started.

10. Activity 12 is completed at the beginning of month 59 and there are no new activities that can start.

11. At the beginning of month 64, activity 5 is completed and activity 6 is started.

12. Activity 9 is completed at the beginning of month 72, leaving a total of 10 men employed for the last two months of the project.

In this case the activities can be scheduled so that the project can be completed within the minimum project duration.

6.5 NETWORK COMPRESSION

The completion of each activity within a project requires a certain quantity of resources and a specific amount of time. If more resources are applied to an activity, it may be possible to complete the activity in less time but at a greater cost. In actual projects a change in the environment may cause the construction work to fall behind schedule.

Assume that the project schedule is the one given in Figure 6.9 and that at the beginning of month 46 there is a labor strike that lasts for 3 months. The engineer must decide which activities should be completed at a faster pace and how much faster in order to complete the project in 73 months. At the beginning of month 46, activity 5 is just ready to start, activity 6 cannot be started until activity 5 is completed, activity 9 is 8/34 completed, activity 11 is 9/10 completed, and activity 12 cannot be started until activity 11 has been completed. The cut-set shown in Figure 6.10 indicates the portion of the project completed at time of delay.

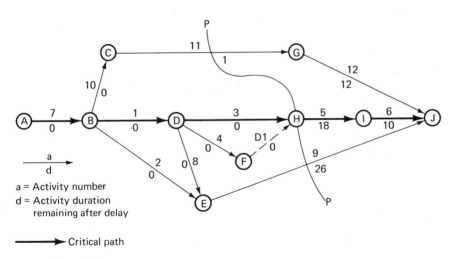

Figure 6.10

Cut-Set Indicating Portion of Project Completed at Time of Delay

All activities to the left of the cut-set have been completed and all activities to the right of the cut-set remain to be completed, including 1/10 of activity 11 and 26/34 of activity 9. The strike causes a construction delay of 3 months and the project must now be completed in 25 work months instead of 28 to be able to meet the 73 calendar-months deadline. In evaluating any changes in schedule, the engineer is interested only in that portion of the graph to the right of the cut-set.

Figure 6.10 indicates that activities 11 and 12 can be completed in 13 months and, therefore, can continue at their normal pace when work is resumed. However, activities 5 and 6 require 28 months, so their duration must be shortened by a total of 3 months. Also, the duration of the remaining portion of activity 9 must be shortened by 1 month. This reduction of the duration is called activity duration compression.

The duration of the activities can be compressed only by committing more resources to the activities than would normally be required. This increases the cost to the contractor and, of course, he wants to find the minimum cost of decreasing the duration of the activities so that he can meet the 73 calendar-months deadline.

A time-cost curve must be developed for each activity that has the potential of being compressed. A typical time-cost curve is shown in Figure 6.11, where C_n is the cost of completing the activity under normal

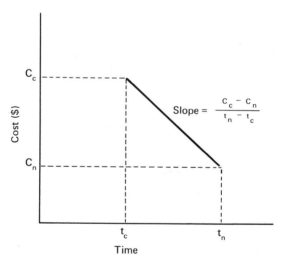

Figure 6.11

Time-cost Curve for a Selected Activity

conditions with a duration of t_n and C_c is the cost of completing the activity under crash conditions in the shortest duration possible of t_c. The slope of the curve is the cost for compressing the activity duration one unit of time. Therefore, the time-cost curve can be defined for each

Table 6.3

TIME-COST DATA FOR ACTIVITIES 5, 6, AND 9

| ACTIVITY | DURATION, MONTHS | | COST, $ | | COST SLOPE |
	Normal	Crash	Normal	Crash	$/month
5	18	16	2.5×10^6	2.7×10^6	$.1 \times 10^6$
6	10	7	6.0×10^6	7.5×10^6	$.5 \times 10^6$
9	26*	22	1.53×10^6	1.73×10^6	$.05 \times 10^6$

*Data for activity 9 is for 26 work months remaining at time of strike.

activity by specifying its normal and crash durations and its normal and crash costs.

Table 6.3 contains the time-cost data for activities 5, 6, and 9.

The project compression calculations are made as follows:

1. List the activities on the critical path; i.e., identify the subgraph of the network model that must be considered. (Activities 5 and 6.)

2. Remove from this list those activities that cannot be compressed because the normal and crash durations are identical or they have already been fully crashed in previous stages. (None.)

3. Select the activity with the smallest cost slope, since this will result in the cheapest compression. (Activity 5.)

4. Determine the amount by which this activity can be compressed. (2 months.)

5. Determine if this compression results in a new critical path. (Activity 9 becomes a critical path also; thus, there are two critical paths.)

6. If a new critical path results, carry out the compression only to the point where the new critical path is the same as the compressed old critical path. (Compress activity 5 by 2 months.)

7. Compute the new project duration and cost. (The remaining project duration is now 26 work months and the remaining project costs are 10.23×10^6.)

8. Steps 1 through 7 are repeated until the desired project duration is achieved or until each critical path through the network model has been compressed to its crash point. (Activities 6 and 9 must both be compressed by 1 month to achieve the 25-months

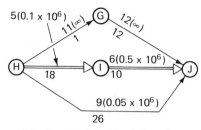

(a) Twenty-eight Month Duration
Remaining After Delay

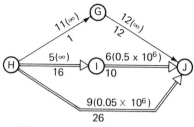

(b) Twenty-six Month Duration
Remaining After Crashing
Activity 5 by Two Months

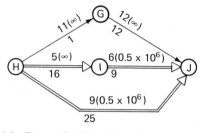

(c) Twenty-five Month Duration Remaining
After Crashing Activity 5 by Two
Months and Activities 6 and 9 by
One Month Each

Figure 6.12

Network Compression of Reservoir Construction

duration because they are on different critical paths. A reduction of 1 month will not result in another new critical path, so the new duration is 25 months at a cost of 10.78×10^6.)

The above computations are illustrated graphically in Figure 6.12.

6.6 A SYSTEM MODEL

The variables associated with determining the earliest start times for a CPM network are activity durations d_{IJ}, float times F_{IJ}, available times between nodes T_{IJ}, and nodal times T_I^E and T_J^E.

The following equation can be developed for each activity in the network model:

$$d_{IJ} + F_{IJ} = T_{IJ} = T_J^E - T_I^E \qquad (6.3)$$

If all the equations associated with the activities are collected together, it is possible to formulate the relationship as a matrix algebra equation, as shown in Equation 6.4. For the garage project of Sections 6.2 and 6.3, and the arrow notation model of Figure 6.4b:

$$\mathbf{D}_{(15\times1)} + \mathbf{F}_{(15\times1)} = \mathbf{T}_{(15\times1)} = \bar{\mathbf{A}}_{(15\times11)}\mathbf{T}'_{(11\times1)} \tag{6.4}$$

where

$\mathbf{D}$ = the column matrix for activity durations (including the zero duration logical dummy activity B–C)

$\mathbf{F}$ = the column matrix for activity float times

$\mathbf{T}$ = the column matrix for activity branch available times

$\mathbf{T}'$ = the column matrix of nodal earliest start times

$\bar{\mathbf{A}}$ = the augmented branch-node incidence matrix of the linear graph CPM model

The combinatorial calculation for the earliest nodal times $\mathbf{T}'$ commence with the known activity duration $\mathbf{D}$ and selectively choose those activities on the earliest time paths so that the nodal time vector $\mathbf{T}'$ can be determined (see Equation 6.1). Once the vector is obtained, the branch available time vector $\mathbf{T}$ can be determined using the relationship provided by the incidence matrix $\bar{\mathbf{A}}$.

The calculations can be written symbolically as follows: Given $\mathbf{D}$ and $\mathbf{A}$ find earliest start tree $\mathbf{B}_E$ so that

$$\mathbf{T}' = \mathbf{B}_E\mathbf{D}$$
$$\mathbf{T} = \bar{\mathbf{A}}\mathbf{T}' \tag{6.5}$$
$$\mathbf{F} = \mathbf{T} - \mathbf{D}$$

where $\mathbf{B}_E$ is a graph structural property for the earliest start tree that can be mapped as a topological matrix using branches as columns and nodes as rows.

Similarly, for latest start node times, the calculations require the selective subtraction of activity duration from the project duration constant using a latest start tree $\mathbf{B}_L$. Again, a symbolic matrix formulation is possible.

There is a similarity in the relationship between the nodal and branch quantities in both organization and physical system problems (see Chapter 3).

Finally, it is possible to formulate the CPM critical path problem as a linear programming mathematical model. Thus, for example, for the garage project and using the arrow notation model, the linear programming model in matrix algebra notation is as follows:

Minimize

$$T_K^E - T_A^E$$

subject to the constraints: (6.6)

$$\mathbf{D} \leq \bar{\mathbf{A}}\mathbf{T'}$$

The specific constraint equation for activity 5 becomes, for example,

$$d_5 \leq T_E^E - T_D^E \tag{6.7}$$

The addition of the float time F_5 for the activity enables the inequality of Equation 6.7 to be replaced by the equation

$$d_5 + F_5 = T_E^E - T_D^E \tag{6.8}$$

so that Equation 6.6 can be rewritten as

Minimize

$$T_K^E - T_A^E$$

subject to:

$$\mathbf{D} + \mathbf{F} = \bar{\mathbf{A}}\mathbf{T'} \tag{6.9}$$

and

$$\mathbf{T'} \geq \mathbf{0}$$

Equation 6.9 shows the typical linear programming model form considered in Chapter 5. It is interesting to note, therefore, that the combinatorial calculation illustrated earlier in this chapter in effect provides a solution to a linear programming problem.

6.7 SUMMARY

It is important to note that throughout the modeling of organizational systems the responsibility for developing the model is placed on the engineer. Thought must first be given to determining events or activities that must be accomplished for project completion. These activities comprise the components of the project. Next, the sequencing or logic of the activities must be determined. This sequencing gives a structure to the problem; thus, it can be modeled in graph form.

Finally, attributes can be assigned to the graph in terms of time or resources. This permits an analysis of the graph as a network. The resulting network model becomes a plan for the project work.

The network model can also be the basis for an analysis procedure to determine the schedule of activities so that the project can be completed efficiently with a limited amount of a particular resource or so that there is a nearly constant rate of resource use.

If the project falls behind schedule, the engineer can use the network model to determine which activities should be completed at a faster pace in order to complete the project within the time specified by the contract.

In construction practice, network analysis is playing an ever-increasing role in project planning and management (see Chapter 10).

The relationship between node and branch quantities for organization networks can be expressed in terms of topological matrices similar to those for physical system problems. And finally, the CPM analysis procedure for minimum project duration is a solution to a linear programming problem.

6.8 PROBLEMS

P6.1. Figure P6.1 shows a CPM network for a project in arrow notation in which the durations are given in number of weeks.

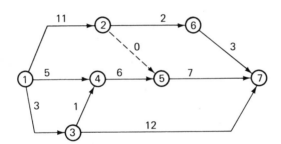

Figure P6.1

1. Compute the following for each job in the schedule: EST, EFT, LST, LFT, TF, FF, and IF.

2. Suppose that all the jobs have been scheduled to start as early as possible and that the works have been on schedule up to the end of week 5. There is a strike on week 6 causing a delay of 1 week. Draw a CPM diagram for the jobs remaining to be done when work resumes on week 7.

P6.2. A project is defined below in tabular form using arrow notation for the various project operations. The table also gives estimated operation times, costs, and crew sizes for the range of feasible rates for working the operations

OPERATION	DURATIONS		CASH $		CREW SIZE
	Normal	Crash	Normal	Crash	
1, 2	6 days	4 days	700	840	6
1, 3	12	10	300	420	7
1, 4	4	2	200	260	5
2, 3	8	6	900	1,000	3
2, 4	4	2	600	760	4
2, 6	15	8	100	380	4
3, 5	6	3	660	960	6
5, 6	2	1	500	600	5
4, 6	6	4	400	500	7

The financial arrangements are such that the indirect cost for overhead will be $50/day for the first 20 days and then will increase by $50/day for each day thereafter. Thus, the indirect cost will be $100 for day 21, $150 for day 22, etc. The type of job allows any man to perform any activity, but once an activity is started it must be completed. That is, the activities cannot be interrupted or segmented.

A complete financial and resource project study is required based on the following analyses:

1. Perform the CPM analysis and develop the bar chart form showing all the TF, FF, and IF values.

2. Draw graphs that show daily resource requirements (number of workers) and cumulative cost for both the EST and LST schedules.

3. Determine the schedule of operations that will allow an orderly increase in men to the peak requirement and then allow an orderly reduction in work force (i.e., try to minimize the fluctuation in the number of men required). What is the schedule and number of men required?

4. Assume that there is a constraint of a maximum of 17 men/day available for the job. Set up the schedule with the shortest completion time subject to this constraint.

5. Obtain the direct, indirect, and total project costs without regard to the resources available, for the normal and possible crash solutions.

6. What is the minimum cost duration?

7. What is the duration and schedule for each activity under (6)?

P6.3. An engineer who has been assigned to supervise construction of a highway wishes to develop a CPM network model as a management aid.

In constructing a highway, the basic sequence of operation is: surveying, clearing and grubbing, moving dirt, placing subbase material and drainage, paving, curing pavement, forming shoulders, seeding embankments with grass, painting lane marks, installing signs, and clearing site. The construction plan calls for highway construction to proceed from one end of the route to the other.

The separate crews for each construction operation are to be organized to progress at the rate of 4 miles per week, the highway being 40 miles long.

Develop a circle notation network model in sufficient detail to indicate logically the construction planning and determine the minimum project duration required in the field. Then translate the model into one using arrow notation. Set out the network roughly to scale, with each horizontal chain representing the progress of one construction operation, such as surveying, clearing, etc.

P6.4. A pump that supplies water for a production process must be replaced. The company would like to replace it during a shift when the plant is not working. The job requires at least two men and a hoist to lift the pump, which weighs 1,000 lb. The following activities have been identified:

1. Shut down pump
2. Close downstream valve
3. Close upstream valve
4. Drain downstream pipe connections
5. Drain upstream pipe connections
6. Remove upstream connections
7. Remove downstream connections
8. Remove nuts from anchor bolts
9. Lift pump and set it aside
10. Position replacement pump
11. Align the replacement pump
12. Tighten nuts on anchor bolts
13. Replace upstream connections
14. Replace downstream connections
15. Prime pump
16. Open upstream valve

17. Open downstream valve

18. Test pump

The pipe is connected to the pump with flange connections. The pipe is 6″ in diameter. Develop a linear graph model of this process and discuss the selection of activities included in the process. Estimate the time required for each activity and determine the time required to replace the pump.

P6.5. In constructing a high-rise building there is a regular progression of activities and building trades from floor to floor. After a certain initializing period, a steady state process and work force develops. The balanced progression of activities, trades, and crews up the building is desirable to project management and is the basis for scheduling crew sizes.

Several activities, trades, and crew sizes for the construction of a typical reinforced concrete brick wall building are:

ACTIVITIES	FLOOR i RELATIONAL LOGIC	ACTIVITY CREW				
		Labor- ers	Car- pen- ters	Iron Work- ers	Brick Ma- sons	Cement Fin- ishers
Erect formwork for floor and columns*	i	1	5	0	0	0
Make and place steel reinforcement cages*	i	0	0	5	0	0
Place concrete for floor and columns*	i	6	3	0	0	5
Dismantle formwork	$i-2$	2	1	0	0	0
Wall bricklaying	$i-3$	6	0	0	10	0
Set windows in walls	$i-4$	1	2	0	0	0

*These activities are concurrent because this building has a large floor plan.

Assuming that each crew finishes its activities on a floor each week:

1. Develop a network model portraying the construction logic for the first 6 floors of the building.

2. Develop a bar chart model and show that a steady state labor work force develops.

3. Indicate those "line of balance" cut-sets that portray the relational logic among the activities.

P6.6. A contractor has been awarded a contract to build a large concrete dam to be used for both flood control and power generation on the Colorado River in southern Nevada. The river is located in a gorge with steep rock sides and bottom. A schematic outline of the dam is shown in Figure P6.6.

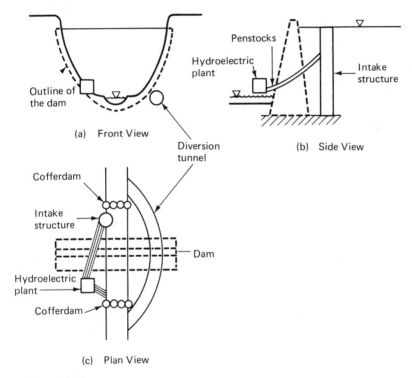

(a) Front View

(b) Side View

(c) Plan View

Figure P6.6

He wishes to develop a construction plan showing the logic and order of the work task activities that must be performed to complete the contract. He has prepared an initial list of activities as follows:

1. Prepare foundation and canyon wall for dam.
2. Place cofferdams to divert water into diversion structure.
3. Place concrete for dam.

4. Construct rock crushing and screening plant.
5. Construct penstocks through the dam to the hydroelectric plant.
6. Construct concrete mix plant.
7. Construct housing for the work force.
8. Construct intake structure at intake of penstocks.
9. Transport labor to the site.
10. Construct hydroelectric plant buildings.
11. Transport equipment to the site.
12. Test performance of dam and hydroelectric plant.
13. Clean up site.
14. Bidding process for turbines.
15. Manufacture and delivery of turbines.
16. Installation of turbines.
17. Construct water diversion structure in wall of gorge.

Develop a suitable construction plan in both circle and arrow notation form. It may be necessary to break down and redefine several activities into smaller work tasks in order to better portray the construction logic.

P6.7. A contractor has received a contract to construct an overpass over an interstate highway. The contract also includes the approach structures (embankments) for the bridge and the concrete slab for a distance of 0.500 mile on each side of the centerline of the bridge. The bridge is to consist of rolled steel beams of standard section with a concrete slab deck. He has prepared an initial list of activities and their work-day durations as shown below:

	Time in *Work Days*
1. Clear site and construct temporary access road.	6
2. Foundation, forms, and reinforcing steel for north abutment.	16
3. Obtain bids for steel girders between abutments.	15
4. Survey for locating improvements and grade stakes.	4
5. Order steel girders.	15
6. Foundation, forms, and reinforcing steel for south abutment.	16
7. Place fill on north side.	45

8. Foundation, forms, and reinforcing steel for center abutment.   16
9. Construct forms for concrete deck over steel girders.   10
10. Place steel girders between abutments.   15
11. Pour concrete deck over steel girders.   4
12. Place fill on south side.   45
13. Place steel railings and other accessories on bridge.   8
14. Pour concrete slab on south side of bridge.   8
15. Prepare subgrade for concrete slab on north side of bridge.   10
16. Pour concrete slab on north side of bridge.   8
17. Finish grading for drainage for entire site.   5
18. Prepare subgrade for concrete slab on south side of bridge.   10
19. Pour concrete for center abutment.   6
20. Place topsoil and sow grass seed for entire site.   4
21. Pour concrete for south abutment.   5
22. Pour concrete for north abutment.   5

The times given for pouring concrete do not include curing times. In general, two days should be provided for curing before any activity is permitted on the concrete.

Develop a feasible construction plan and portray your plan in linear graph form using both arrow and circle notation. Then determine a bar chart model for the project indicating total, free, and interfering floats. Define a project schedule based on an earliest start policy.

P6.8. Give reasons why you agree or disagree with the following:

1. CPM networks use unidirectional concepts to model project logic; i.e., no cyclic paths can exist.

2. CPM networks are useful for situations involving many activities and heavily interfacing logic.

3. CPM networks are not useful for heavily repetitive projects such as pipe laying unless gross modeling concepts are used.

4. A trade-off must exist in any model between the accuracy of the modeling logic and representation and the usefulness of the model to the decision-maker.

5. Projects with cyclic logic are best modeled with flow and/or simulation models.

6. Different CPM models are possible for the same project depending on the user and his purpose.

7

DECISION ANALYSIS

The Little Miami facility is one of three sewage-treatment plants built in the city of Cincinnati to curb pollution of the Ohio River and is a part of the Ohio River Valley Clean Streams Program that was selected as the Outstanding Civil Engineering Achievement of 1963. The Ohio River Valley Program involves eight states, over 1,500 municipalities, and hundreds of industrial facilities adjacent to the 1,000 miles of waterways. Over 1,200 industrial plants have installed pollution control facilities that have been rated acceptable by the control agencies. (Courtesy of the American Society of Civil Engineers)

7.1 ELEMENTS OF A DECISION PROBLEM

The primary objective in a decision problem is to choose the optimum plan or policy from a specific set of possible alternatives. However, the decision must be made in a logical and subjective manner so that all possible alternatives and their consequences are considered and the final decision can be explained and justified to the interested parties. In addition, contingency plans must also be prepared for all potential consequences of a decision.

For example, consider the decision problem of the contractor discussed in Example 2.4 of Chapter 2. The contractor must choose among three alternatives:

1. To move the equipment away from the river bank to avoid loss due to potential flooding;
2. To leave the equipment at the location and build a protective platform; or
3. To leave the equipment at the location and not build a protective platform.

Each alternative results in a certain cost to the contractor, and the cost of the last two alternatives will depend on the extent of flooding during the spring. Therefore, in making his decision, the contractor must consider the chances of flooding and the potential loss that may result. Furthermore, should he decide to leave the equipment on location, he must have available contingency plans for replacing the equipment in case of serious loss from a flood.

Two common elements characterize all engineering decision problems:

1. Probabilistic events; and
2. Insufficient data.

Probabilistic events are occurrences that are beyond the control of the decision-maker, although the probability or percentage chance of these occurrences can usually be predicted from either historical data or engineering foresight. For example, inclement weather is beyond the control of the engineer who is planning a construction schedule, yet it is a factor that he must take into consideration. Natural events such as flooding, earthquakes, and hurricanes are vital factors that must be considered in many engineering decision problems, including the structural design for high-rise buildings, location of dams and reservoirs, and location of transportation systems.

The amount of technical data that is available to a decision-maker is usually limited by many factors. Among the more common factors are: technical skill of the decision-maker, state of the art in science and technology, limited financial resources, limited research capability, and limited time available for research and experimentation. Although technical data can usually be obtained at a certain cost to the decision-maker, the value of the additional information must be weighed against the cost of obtaining such information. For example, traffic surveys may be conducted to provide data on the volume of traffic flow, degree of congestion, commuter traffic behavior, percentage of truck and through traffic, and so on. Such surveys require time and money but provide in return information that traffic engineers may use to make sound decisions on the design of city traffic systems.

A decision problem involves a set of alternative actions that are connected in some way to a set of possible outcomes. If the decision-maker knew which outcomes would occur with each act, he could choose the act that resulted in the outcome he most valued. Actions may lead to outcomes where other decisions must be made and many actions lead to outcomes that depend on chance; and, therefore, the outcome is known only in terms of probability of occurrence if certain actions are taken. The necessary first step in the decision-making process, therefore, is to identify the alternative actions and outcomes and their interrelationships. The value of each outcome and the probabilities of its occurrence for the alternative actions should be quantified. These parameters, along with their interrelationships, constitute the problem model, which may then be analyzed to determine the set of actions the decision-maker should take to achieve his most valued outcome.

7.2 THE DECISION MODEL

Example 2.4 in Chapter 2 has already illustrated the usefulness of linear graphs in modeling a decision problem. The graphical model, called the decision tree, delineates all the alternatives and their possible consequences. It helps to facilitate the task of decision analysis. To further illustrate the purpose and technique of constructing the decision tree, consider the following problem.

Problem Statement

A contractor working on an outdoor construction project in a coastal area is reviewing his progress on August 1. He finds that if he keeps the normal speed and loses no time because of hurricanes, he will be able to complete the job on August 31. However, due to the poor weather conditions in the area after August 16, he will have only a 40 percent chance of finishing on time. He estimates that there is a 50 percent chance of a minor hurricane, which will cause a delay of 5 days, and a 10 percent chance of a major hurricane, which will cause a delay of 10 days. He has to decide now whether he should start a crash program on August 2 at an additional cost of $75 per day and finish the project on August 16. As an alternative, he can maintain the normal schedule and review his progress on August 31. At that time, if a hurricane has occurred and the project is delayed, he will have the choice of accepting the delay at a certain penalty cost or he can try to crash the program then. The penalty cost for delay of completion will be $400 per day for the first 5 days and $600 per day for the second 5 days. The additional cost of a crash program after the hurricane will be $200 per day. The total additional cost is computed as the sum of delay penalty cost and crash cost.

He further estimates that the possible results (outcomes) of a crash program after a minor hurricane causes a 5-day delay will be as follows:

CRASH PROGRAM RESULT	PROBABILITY	TOTAL ADDITIONAL COST
Save 1 day	0.5	$1,600 + 800 = $2,400
Save 2 days	0.3	$1,200 + 600 = $1,800
Save 3 days	0.2	$ 800 + 400 = $1,200

The possible results of a crash program after a major hurricane causes a 10-day delay is estimated as follows:

CRASH PROGRAM RESULT	PROBABILITY	TOTAL ADDITIONAL COST
Save 2 days	0.7	$(2,000 + 1,800) + 1,600 = 5,400
Save 3 days	0.2	$(2,000 + 1,200) + 1,400 = 4,600
Save 4 days	0.1	$(2,000 + 600) + 1,200 = 3,800

The above problem description adequately outlines all the alternatives, the consequences, as well as all the relevant data needed to make the decision. Therefore, this description constitutes a descriptive model of the decision problem. It is obvious, however, that even for a problem as simple as this, it is difficult to obtain an overall view of the problem from such a descriptive model. As an interesting experiment, before reading any further, assume that you are the contractor and try to solve the above problem.

Model Formulation

Figure 7.1 illustrates the different steps used to construct the decision tree in the above problem. These steps are described below.

1. A decision node (□) is drawn to represent the most immediate decision that the contractor must make. A branch is drawn from the node to represent each alternative that is available to him at this decision point, as shown in Figure 7.1a.

2. The potential results of each of the decision alternatives are then modeled. The outcome of immediately launching a crash program is that the project will be completed at an added cost of $75 per day for 15 days, or a total of $1,125. This is represented by assigning a terminal node (Δ) with a value of −$1,125, as shown in Figure 7.1b. The second alternative has three possible outcomes: no delay, 5-day delay, and 10-day delay with a probability of 0.4, 0.5 and 0.1, respectively. This is represented as a chance node (O) where each possible result is represented by a branch and the probability of occurrence is entered along the branch.

3. The only possible result of no delay on August 31 is that the project is completed on time with no delay cost. This is represented by a terminal node with a value of −$0.0. For the other two outcomes, the contractor must decide next whether to launch a crash program or to maintain a normal pace. Thus, each is modeled by a decision node with branches issuing from the node to represent the possible alternatives, as shown in Figure 7.1c.

4. The decision tree is finally completed by again using a branch to represent each possible outcome, and the total cost resulting from that outcome is written next to the terminal nodes, as shown in Figure 7.1d.

Constructing the decision tree, however, is only the last and perhaps the easiest stage in modeling a decision problem. The greatest challenge lies in identifying and stating the sequential decisions and their possible consequences that constitute the problem model. Estimating costs and benefits and assigning probabilities to the chance occurrences are tasks that test the skills of the decision-maker, and the accuracy of these estimates ultimately control the validity of the final decision.

The concept that any problem model must of necessity be an abstraction of reality is also well illustrated by the above example. Although the exact delay caused by a hurricane may range anywhere from

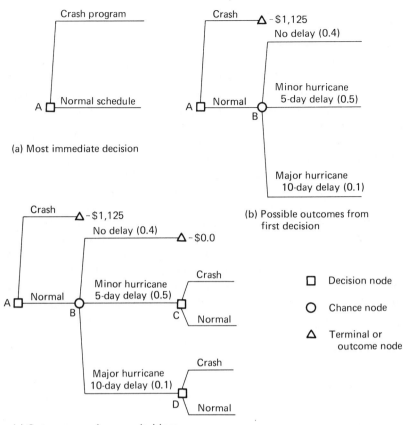

(a) Most immediate decision

(b) Possible outcomes from first decision

☐ Decision node

◯ Chance node

△ Terminal or outcome node

(c) Outcomes require more decisions

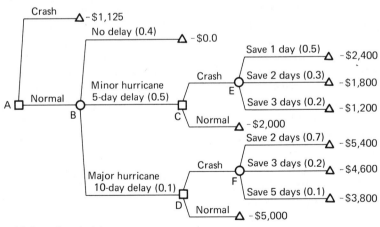

(d) Complete decision tree

Figure 7.1

Development of Decision Tree Model of Contractor's Problem

one day to ten days or more, the contractor has simplified the problem by considering only the two most likely cases—a minor hurricane resulting in a 5-day delay, or a major hurricane causing a 10-day delay. This abstraction is easily justifiable if the nature of the construction job is such that delays must be counted in intervals of several days; jobs such as pouring and drying concrete and exterior painting usually need several days of continuous good weather. In addition, considerations must be given to the repair of hurricane damages and to the loss in momentum due to the delay.

On the other hand, delays of any discrete number of days can be built into the decision model. The amount of time that can be saved by the crash programs may also be counted in half-day intervals. In reality, the contractor may also have the option of keeping his program under continuous review depending on the progress between August 1 and 16. Indeed, the model can be expanded to include some or all of these alternatives, with a resulting increase in the complexity of the model. However, in order to keep the problem on a manageable scale for the purpose of analysis, some degree of abstraction must be exercised using engineering judgment.

7.3 FUNDAMENTALS OF PROBABILITY

It is beyond the scope of this book to present a detailed discussion of the theories of probability. However, since probability is such an integral part of the method of decision analysis, it is useful to review here some of the basic fundamentals of the subject.

The probability that a chance event will occur is measured in a continuous scale ranging between 0 and 1. An event that has absolutely no chance of occurring has a probability value of zero. On the other hand, an event that is sure to occur has a probability value of 1. Thus, any event that has a chance of occurring will have a probability value between 0 and 1. The more likely it is to occur, the higher will be its probability value. In symbolic language, the probability that an event A will occur is denoted as $P(A)$.

In the example discussed in Section 7.2, the following probabilities were cited for the three possible results in case the contractor decided to maintain construction at his normal pace.

$$P(\text{no delay}) \quad\quad = 0.4$$
$$P(\text{minor hurricane}) = 0.5$$
$$P(\text{major hurricane}) = 0.1$$

These probability values provide a measure of the relative chance of these three events occurring. They can be determined in many ways. They may simply be based on an educated guess by the contractor, who bases his judgment on present weather conditions and on his recollection of weather conditions at the same period in previous years. They may also be the result of a statistical analysis of the historical records of local weather conditions. Suppose that the past 50 years of weather records for the area are available to the contractor. He can count the number of years in which there was either a minor or major hurricane during the period of August 16–31. The above probabilities can then be computed as follows:

$$P(\text{no delay}) = \frac{\text{number of years with no hurricane}}{\text{total number of years counted}} = \frac{20}{50} = 0.4$$

$$P(\text{minor hurricane}) = \frac{\text{number of years with a minor hurricane}}{\text{total number of years counted}} = \frac{25}{50}$$

$$= 0.5$$

$$P(\text{major hurricane}) = \frac{\text{number of years with a major hurricane}}{\text{total number of years counted}} = \frac{5}{50}$$

$$= 0.1$$

The greater the number of years of weather records used in the analysis, the more reliable will be the computed probability value. However, it is important to realize that these probability values indicate the chance occurrence of these events during the *past* years. In projecting these figures for the present, the contractor is assuming that past weather conditions are a good indicator of what is going to occur this year. A more realistic set of probability values for the above events may possibly be determined from a meteorological study of the existing weather conditions and from long-range weather forecasts.

The above methods apply equally well to the occurrence of other uncertain events such as earthquakes, floods, flood damages, stream flows, traffic accidents, bearing strength of the soil, and the exact tensile strength of a concrete or steel beam. Figure 7.2 is a typical frequency histogram showing, for example, the number of rainy days in August for the past 50 years. Thus, the probability that there will be x number of rainy days in this August can be estimated as follows:

$$P(x \text{ rainy days}) = \frac{\text{number of years with } x \text{ rainy days}}{\text{total number of years counted}}$$

In many instances, the chance variable may have a value within a continuous range rather than a discrete set of possible values. For example,

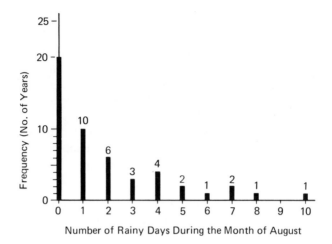

Figure 7.2

Frequency Histogram

the exact tensile strength of a given steel beam may have a value between 1.00 kip/in.² and 1.15 kip/in.² In such cases, the pattern of occurrences may be modeled by a mathematical function, $f(x)$, which is called the probability density function of that variable. Then the probability that the variable x may take on a value between, say, x_1 and x_2 can be computed by integration as follows:

$$P(x_1 \le X \le x_2) = \int_{x_1}^{x_2} f(x) \, dx$$

Figure 7.3 illustrates such a probability model for the tensile strength of a steel beam.

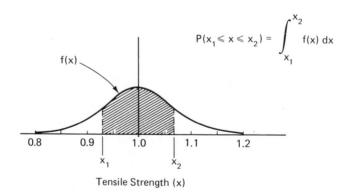

Figure 7.3

Probability Density Function

Another basic property of probability may be learned from the above problem concerning the contractor. It has been assumed that there is no chance that more than 10 days can be lost due to inclement weather. That is,

$$P(\text{delay longer than 10 days}) = 0$$

In fact, it has been definitely assumed that there is to be either no delay, a delay of 5 days, or a delay of 10 days. That is,

$$P(\text{either no delay, a delay of 5 days, or a delay of 10 days}) = 1$$

Therefore, the summation of the probability values of the above three mutually exclusive (no two events can occur simultaneously) events must be equal to 1. In general, let x_1, x_2, x_3, . . . , x_n be n mutually exclusive events that could possibly occur at a chance node in the decision tree, and let $P(x_i)$ denote the probability that event x_i occurs; then

$$\sum_{i=1}^{n} P(x_i) = 1 \qquad (7.1)$$

7.4 DECISION ANALYSIS BASED ON EXPECTED MONETARY VALUE

Consider the problem of the contractor when he is evaluating the course of action he would take on the assumption that he is at the decision node C. This portion of the tree is redrawn in Figure 7.4 for easy reference.

The contractor is now already 5 days behind schedule, and he has to decide whether to launch a crash program immediately or to keep the normal pace. If he decides to do the latter, he stands to lose $2,000 for sure. If he decides to crash, there is a probability of 0.3 that he would lose only $1,800 and a probability of 0.2 that he would lose only $1,200.

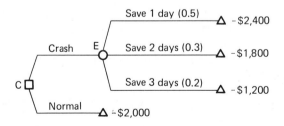

Figure 7.4

Decision Tree Confronted by Contractor If He Is 5 Days Behind Schedule on August 31

However, there is a probability of 0.5 that he would lose as much as $2,400. Which alternative he would choose will, of course, depend on the value he places on money and on his willingness to take chances. But he must make his decision based on the comparative value of the two choices before he knows the final outcome that will actually occur.

Expected Monetary Value

One way of comparing the values of alternatives is to compare their expected monetary values. In the present example, the expected monetary value of the choice to crash is computed as follows:

Expected Monetary Value (EMV) at node E = 0.5 × (−$2,400)

+ 0.3 × (−$1,800) + 0.2 × (−$1,200) = −$1,980

It means that if the contractor has the opportunity to make the same decision in many other jobs under similar circumstances, he would lose on the average $1,980 by choosing to crash. However, if he chose to keep the normal pace every time, he would lose $2,000 each time. Thus, if the decision-maker is a believer in EMV, then his logical choice is to start a crash program.

In general, let $x_1, x_2, x_3, \ldots, x_n$ be the monetary value associated with each of the n branches at a chance node and $P(x_1), P(x_2), P(x_3), \ldots, P(x_n)$ be the corresponding probabilities. Then the expected monetary value (EMV) of that chance node is defined as follows:

$$\text{Expected Monetary Value} = \sum_{i=1}^{n} x_i P(x_i) \qquad (7.2)$$

Comparative Analysis of Alternatives

The decision problem modeled by the decision tree in Figure 7.1 can be solved by successively comparing the EMV of the alternatives at each decision node. Starting from the tips of the decision tree, the EMV of the chance nodes E and F are first computed. That is,

EMV at node E = −$1,980 (from above)

EMV at node F = 0.7 × (−$5,400) + 0.2 × (−$4,600) +

0.1 × (−$3,800) = −$5,080

These EMV's are then entered into the decision tree at the respective node, as shown in Figure 7.5. By comparing the EMV of the two alterna-

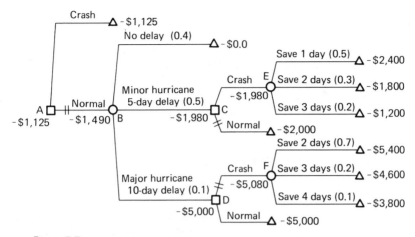

Figure 7.5

Decision Analysis Based on EMV

tives at node C, the best decision is to crash the project. A hatch mark can be drawn on the rejected alternative to denote this choice. Thus, should the contractor have to make the decision at node C, he would choose to crash the project with an EMV of $-\$1,980$. This is equivalent to saying that the decision node C has an EMV of $-\$1,980$. Similarly, should the contractor have to make the decision at node D, he would choose to keep the normal pace with an EMV of $-\$5,000$. Again, a hatch mark is used to cross out the rejected alternative and an EMV of $-\$5,000$ is assigned to node D.

Proceeding one step further, the EMV at chance node B can next be computed as follows:

$$\text{EMV at node } B = 0.4 \times (-\$0.) + 0.5$$
$$\times (-\$1,980) + 0.1 \times (-\$5,000) = -\$1,490$$

Finally, at decision node A, the alternative to crash imposes a certain loss of $\$1,125$, while to keep the normal pace would lead to an EMV of $-\$1,490$. Thus, the better choice is to crash the program immediately.

In general, the analysis procedure consists of the following steps:

1. Starting from the tips of the tree, compute the EMV at the chance nodes closest to the tips;

2. At each decision node, compare the EMV's of the alternatives and choose the alternative with highest profit or minimum loss. Assign the EMV of the chosen alternative to that decision node.

3. Proceed node-by-node toward the root of the tree. The EMV for the decision problem is obtained at the root of the tree.

4. Trace back through the tree to determine the optimum decision as indicated by the branches that do not have hatch marks. This set of decisions constitutes the optimum strategy. In this case, there is only one decision: crash the project.

Implications of the EMV Criterion

In using the EMV as a decision criterion, a decision-maker must always keep in mind that this approach carries the following two implications:

1. The decision-maker is betting on the law of averages, since the EMV of an alternative means that if he chooses this alternative many times under similar conditions he would receive this much in return on the average. This may be quite different from the amount that is actually obtained if the choice is only made once because only one final outcome results.

2. The EMV is a completely objective measure of the value of money and implies that every dollar within a sum of money provides the same amount of satisfaction. It does not consider personal differences about the value of money.

For example, consider the decision problem illustrated in Figure 7.6. Alternative A has a 30 percent chance of making $100,000 profit and a 70 percent chance of losing $20,000. Alternative B has a 50 percent chance of making a $5,000 profit and an equal chance of gaining $6,000. According to a strict EMV analysis, alternative A has an EMV of $16,000 and alternative B has an EMV of only $5,500; hence, alternative A is the better choice. On the other hand, it is obvious that in practice many people (including all the authors) would be quite happy to accept alternative B but absolutely refuse to accept alternative A. However, alternative A may be the preferred choice of one who draws little satisfaction from the meager sum of $20,000.

In reality, a decision-maker often bases his financial decisions on his available capital and on his willingness to take risk. He usually tries to prevent a total loss of his capital, but he must be willing to risk some

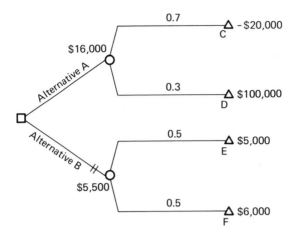

Figure 7.6

Decision Problem for Which EMV May Not Be a Suitable Criterion

loss for the opportunity to profit. For example, consider a young, aggressive investor who has a capital of $50,000. Confronted with the above decision problem, he would probably be strongly influenced by the opportunity of gaining $100,000 and thus tripling his capital. The prospect of losing $20,000 would not seriously deter him from taking alternative A, because even if he loses $20,000, he still has $30,000 remaining as capital. Therefore, the computed EMV of $16,000 for alternative A and $5,500 for alternative B may appropriately indicate the relative values of the two alternatives to this aggressive investor.

Suppose next that the decision-maker is a young, conservative investor who has a capital of only $10,000. Although he would undoubtedly be interested in gaining $100,000, he would also be strongly deterred by the prospect of losing $20,000. It would not only wipe out his capital, but would put him $10,000 in debt. Since alternative B means a certain gain of either $5,000 or $6,000, which is more than 50 percent of his capital, he would be strongly inclined to choose alternative B. To this investor, the computed EMV's fail to truly measure the relative values of the two alternatives.

The EMV criterion failed in this example because the monetary values assigned to the various outcomes at the terminal nodes C, D, E, and F do not truly reflect the values of these outcomes to the decision-maker. A loss of $20,000 provides different degrees of satisfaction (or dissatisfaction) to different persons depending on their capital and risk behavior.

7.5 DECISION ANALYSIS BASED ON UTILITY VALUE

In order to provide a personal measure of the relative worth of the decision results, the monetary values may be transformed to their equivalent utility values by a utility function. Each decision-maker has his own utility function, and it must reflect his risk behavior and his outlook toward money. For example, let the utility function in Figure 7.7a represent the risk behavior of the conservative investor in the pre-

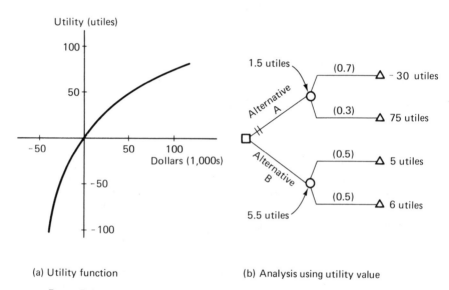

(a) Utility function   (b) Analysis using utility value

Figure 7.7

Decision Analysis Using Expected Utility Value

ceding example. A loss of $20,000 has an equivalent utility value of -30 utiles, and a gain of $100,000, $5,000, and $6,000 has an equivalent utility value of 75, 5, and 6 utiles, respectively. Utiles are defined as units that express personal values. By substituting these utility values for their corresponding monetary values in the decision tree, the expected utility values for alternatives A and B can now be computed as follows:

Expected utility value of alternative $A = (0.7) \cdot (-30) + (0.3) \cdot (75)$

$$= 1.5 \text{ utiles}$$

Expected utility value of alternative $B = (0.5) \cdot (5) + (0.5) \cdot (6)$

$$= 5.5 \text{ utiles}$$

The results are illustrated in Figure 7.7b. Thus, based on the expected utility values, alternative B is the better choice for this conservative investor.

It is no easy matter to develop an accurate utility function for an individual. His risk behavior and his value of money must be thoroughly understood. One method is to test his preferences under a wide range of risk situations involving different amounts of monetary values. This method is best illustrated by considering an example. The following is a step-by-step description of how a utility curve can be established for the contractor's decision problem in Figure 7.5:

1. Identify the upper and lower limits of the monetary values involved in the decision problem and arbitrarily assign utility values to two different monetary values within these limits; e.g.,

	MONETARY VALUES	ARBITRARILY ASSIGNED UTILITY VALUE
Upper limit	$0.0	0
Lower limit	−$6,000	−100

These figures establish two points on the utility curve. Since utility values provide relative measures, these two points merely establish a reference datum for the utility values.

2. Create a hypothetical decision problem that has two possible alternatives and three possible outcomes. Two of the outcomes should be assigned the two monetary values to which utility values have already been given. Figure 7.8a presents such a hypothetical problem. In alternative A, the contractor has a 70 percent chance of losing $6,000 (−100 utiles) and a 30 percent chance of losing nothing (0 utiles). In alternative B, the contractor can purchase insurance to cover any potential loss. What is the maximum amount that he would be willing to pay to cover the 70 percent chance that he might lose $6,000? Suppose that his answer is $5,400. The contractor's answer means, in essence, that the two alternatives have equal worth if alternative B is to lose −$5,400; i.e., the utility value of −$5,400 is equal to the expected utility value of alternative A. Hence,

$$\text{Utility value of } -\$5,400 = 0.7(-100) + 0.3(0) = -70$$

Thus, a third point is established on the utility curve.

3. Create another hypothetical problem that also has two alternatives and three possible outcomes, and two of these outcomes have monetary values of which the utility values are already known. By using the

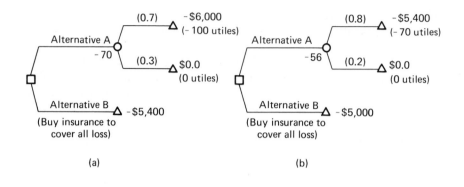

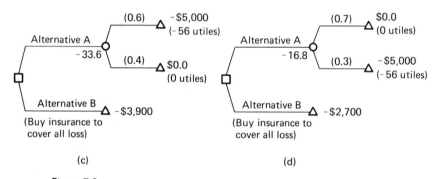

Figure 7.8

Hypothetical Decision Problems

same procedure as above, a new point can be established. This procedure is repeated until enough points are available to establish the utility curve. (See Figures 7.8b to 7.8d.)

4. Finally, a best-fitting curve is drawn to represent the utility function, as shown in Figure 7.9.

Having established the utility curve, the equivalent utility for all the possible outcomes in the decision problem can be obtained directly from the curve. The alternatives are then evaluated according to their expected utility values. Figure 7.10 shows an analysis of the contractor's problem using utility values from the curve in Figure 7.9. If this utility curve truly represents the contractor's risk behavior, his proper choice would be to crash the project. This is the same conclusion reached by EMV in Figure 7.5. However, the decision at node C would be changed. Hence, the results obtained using utility values may be, but are not necessarily, different from those obtained using EMV. The shape of the

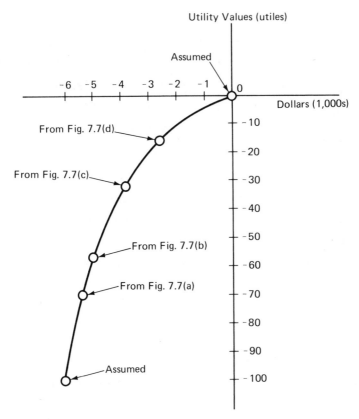

Figure 7.9

Utility Curve for the Contractor

utility curve has a pronounced effect on the results when using utility values.

The above procedure for establishing the utility function can be easily modified for decision problems involving profits instead of losses. For example, the hypothetical problem in Figure 7.11 may be confronted by an individual. In alternative A, he has a 70 percent chance of making a \$50,000 profit and a 30 percent chance of making a \$5,000 profit. Suppose that the utility values for \$50,000 and \$5,000 profits have already been known to be 100 and 20, respectively. In alternative B, the individual has the chance to sell alternative A to another person. What is the minimum amount for which he would be willing to sell A? Suppose that his answer is \$30,000. The utility value of \$30,000 is thus computed as $(0.7) \cdot (100) + (0.3) \cdot (20) = 76$.

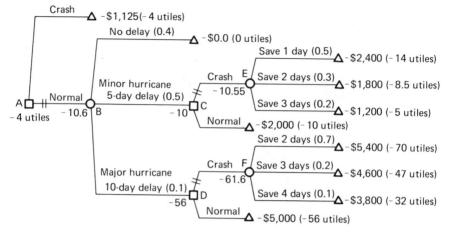

Figure 7.10

Analysis of Construction Problem Using Expected Utility Value

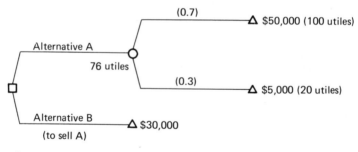

Figure 7.11

Hypothetical Decision Problem to Aid in Establishing a Utility Function Involving Profits

7.6 BAYES' THEOREM

Let A and B be two events such that the occurrence of one event influences directly the occurrence of the other event. Then the conditional probability that event A would occur after knowing that event B has already occurred is denoted as $P(A/B)$, which reads the probability of A given B. Bayes' theorem states that

$$P(A/B) = \frac{P(B/A) \cdot P(A)}{P(B)} \qquad (7.3)$$

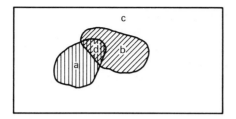

Figure 7.12

Venn Diagram for Events A and B

This theorem can easily be proved using the diagram in Figure 7.12. Let area c bounded by the rectangle represent the sample space, area a the number of occurrences of event A, area b the number of occurrences of event B, and the common area d represent the event that both A and B occur together. Thus,

$$P(A) = \frac{a}{c},$$

$$P(B) = \frac{b}{c},$$

$$P(A/B) = \frac{d}{b},$$

and

$$P(B/A) = \frac{d}{a}$$

Also,

$$\frac{P(B/A) \cdot P(A)}{P(B)} = \frac{\dfrac{d}{a} \cdot \dfrac{a}{c}}{\dfrac{b}{c}} = \frac{d}{b}$$

Therefore,

$$P(A/B) = \frac{P(B/A) \cdot P(A)}{P(B)}$$

This theorem is used extensively in decision analysis for computing the conditional probability of chance events. It is especially useful for problems in which engineering tests are conducted to predict the state of uncertain variables; for example, such as problems in which sounding is used to determine the depth of water, drilling to determine the depth of bedrock, or field testing to determine the strength of concrete. These tests usually cannot reveal the true state of the uncertain variables. The results can, at best, provide more reliable estimates of the probability of occurrence of the various possible states of the variables.

As an example, suppose that the contractor in the earlier problem decides to engage the consulting service of an expert meteorologist whose ability to predict the true state of events is given in Table 7.1. If the

Table 7.1

CONDITIONAL PROBABILITIES $P(S_i/T_j)$

	TRUE STATE		
PREDICTED STATE	T_1 No Hurricane	T_2 Minor Hurricane	T_3 Major Hurricane
S_1 No Hurricane	0.8	0.3	0.1
S_2 Minor Hurricane	0.2	0.6	0.2
S_3 Major Hurricane	0.0	0.1	0.7

true state is T_1, then there is an 80 percent chance that he would predict S_1 and a 20 percent chance that he would predict S_2. Similarly, if the true state is T_2, then there is a 30 percent, 60 percent, and 10 percent chance that he would predict the states S_1, S_2, and S_3, respectively. If the true state is T_3, then the probabilities are 10 percent, 20 percent, and 70 percent for predicting states S_1, S_2, and S_3, respectively.

Figure 7.13 is a decision tree of the problem now facing the contractor. He wants to base his decision on the meteorologist's prediction. Before analysis can begin, the probability must be determined for each of the chance events. Let the conditional probability that T_j is the event that actually occurs after S_i has been predicted be denoted as $P(T_j/S_i)$. It is intuitively obvious that $P(T_1/S_1) \neq P(T_1/S_2)$, and so on. The immediate problem therefore, is to determine these conditional probabilities from the available information in Table 7.1.

The probability $P(S_i)$ for $i = 1$ to 3, can first be computed by enumerating all the different conditions under which the meteorologist would predict state S_i and summing the probabilities of these conditions. Figure 7.14 presents the graph model of all the different combinations of

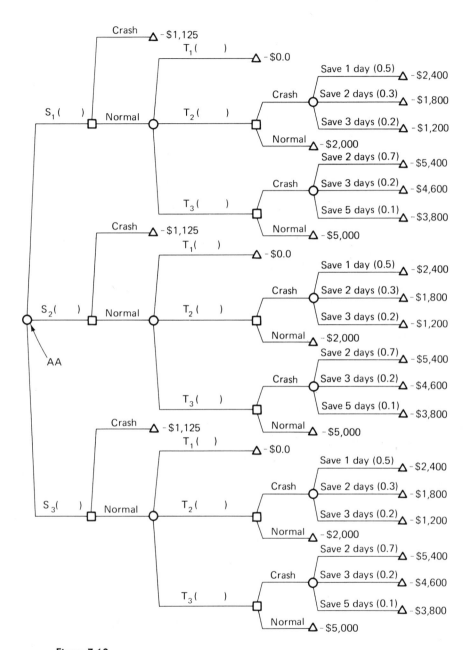

Figure 7.13

Decision Tree for Contractor after Expert Has Been Hired but Before He Has Made His Prediction

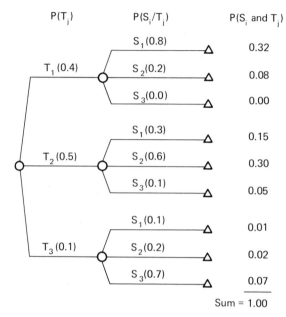

Figure 7.14

Graph Model of Possible Combinations of S_i and T_j

events. The probability that a true state T_j occurs remains the same as in the preceding problem. The probability that the true state is T_j and that the meteorologist predicts S_i, $P(S_i$ and $T_j)$, is then computed as $P(T_j) \times P(S_i/T_j)$. Thus, the probability of each possible combination of S_i and T_j is computed and given in Figure 7.14 in the column $P(S_i$ and $T_j)$. As a check, the sum of the probabilities under this column must be 1. The probability that the meteorologist would predict S_i is then simply computed as follows:

$$P(S_i) = P(S_i \text{ and } T_1) + P(S_i \text{ and } T_2) + P(S_i \text{ and } T_3)$$

Hence,

$$P(S_1) = 0.32 + 0.15 + 0.01 = 0.48$$
$$P(S_2) = 0.08 + 0.30 + 0.02 = 0.40$$
$$P(S_3) = 0.0 \ + 0.05 + 0.07 = 0.12$$

Next, the conditional probabilities $P(T_j/S_i)$ for $i = 1$ to 3 and $j = 1$ to 3 can be computed directly using Bayes' theorem.

$$P(T_1/S_1) = \frac{P(S_1/T_1) \cdot P(T_1)}{P(S_1)} = \frac{(0.8) \cdot (0.4)}{0.48} = 0.667$$

$$P(T_2/S_1) = \frac{(0.3) \cdot (0.5)}{0.48} = 0.312$$

$$P(T_3/S_1) = \frac{(0.1) \cdot (0.1)}{0.48} = 0.021$$

$$P(T_1/S_2) = \frac{(0.2) \cdot (0.4)}{0.40} = 0.20$$

$$P(T_2/S_2) = \frac{(0.6) \cdot (0.5)}{0.40} = 0.75$$

$$P(T_3/S_2) = \frac{(0.2) \cdot (0.1)}{0.40} = 0.05$$

$$P(T_1/S_3) = \frac{(0.0) \cdot (0.4)}{0.12} = 0.0$$

$$P(T_2/S_3) = \frac{(0.1) \cdot (0.5)}{0.12} = 0.417$$

$$P(T_3/S_3) = \frac{(0.7) \cdot (0.1)}{0.12} = 0.583$$

These conditional probabilities can now be entered into the decision tree and decision analysis can next proceed according to the procedures described previously to obtain an expected value of $-\$931.92$, as shown in Figure 7.15.

7.7 VALUE OF INFORMATION

It has been assumed until now that the probabilities and values of the outcomes, either in monetary or utility terms, have been known for the decision problem. With regard to practical problems, it is often necessary to conduct studies or investigations to determine these probabilities and values. The question then arises as to how much should be spent to obtain this information.

Recall the contractor problem from the previous sections. The contractor can hire a meteorologist to obtain a more reliable prediction of weather conditions. How much should the contractor be willing to pay the meteorologist? One way of viewing the problem is to put Figures 7.5 and 7.15 together, as shown in Figure 7.16, where the heavy mark on branch a can be considered as a toll gate for which the contractor

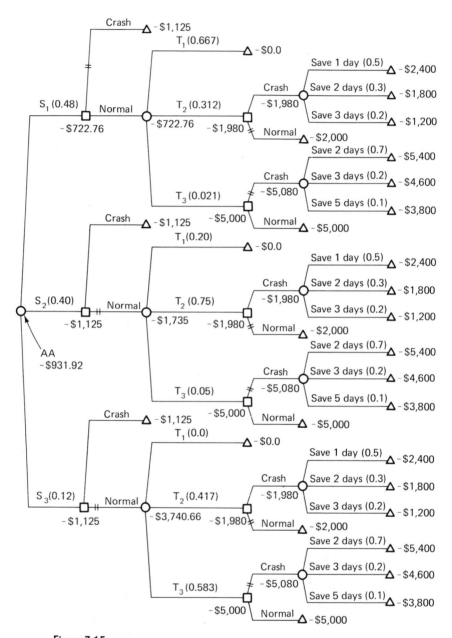

Figure 7.15

Decision Analysis after Expert Has Been Hired but Before He Makes His Prediction

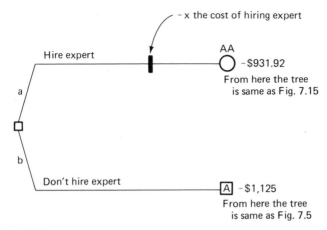

Figure 7.16

Decision Tree to Determine the Value of New Information

must pay a toll if he chooses this alternative. The problem is then to determine the maximum toll the contractor would be willing to pay and still choose to hire the expert. Let this toll be represented by x. If the contractor were indifferent between the two alternatives, his expected value of both would be equal. The expected value after hiring the expert is $(-x - 931.92)$ and the expected value before hiring the expert is $-\$1,125$. Hence,

$$-x - \$931.92 = -\$1,125$$

and

$$x = \$1,125 - \$931.92 = \$193.08$$

which is the upper limit the contractor should pay for the services of the meteorologist. Thus, the value of sampling or investigating to obtain new information is equal to the increase in the expected value that will be derived from this information. In this case, the cost of hiring the meteorologist was considered after the expected values had been computed. Another way to consider this cost would be to add this amount to the cost associated with each outcome that is a possible result of the decision to hire the meteorologist. That is, the cost x would be added to each possible outcome in Figure 7.15 and then the expected value would be computed. The result is the same if the contractor is an EMV'er. However, if the contractor is a non-EMV'er, then all the computations must be repeated using utility values at the terminal nodes.

7.8 SUMMARY

This chapter presents a method for choosing the optimum plan or policy from a specific set of possible alternatives.

The expected monetary value is not always a good criterion because of the personal values of the decision maker which may be quite different from person to person, may vary with time, and is, in general, a nonlinear function of the monetary values. These personal values are indicated by the utility function which is a powerful concept, but one that is difficult to assess precisely.

The approach to decision-making in probabilistic situations should include:

1. Determining the possible results of each of the alternatives;

2. Determining the probability of each outcome;

3. Determining the utility curve for the decision-maker and assigning utility values to the results; and

4. Calculating the expected utility values of the alternatives.

The selection criterion is the maximum expected utility value. A decision tree, a special type of linear graph, can be used to graphically represent the decision problem.

Bayes' theorem provides an easy means for computing conditional probabilities. The decision-maker should continue to acquire new information as long as the increase in expected utility attributable to this new information is greater than the cost.

If it is not possible to estimate the probabilities of the outcomes, then expected values cannot be used as a criterion in decision-making. However, if the problem is important enough to receive formal analysis, the decision-maker must either make a subjective estimate of the probabilities or obtain data that will permit an estimate to be made.

7.9 PROBLEMS

P7.1. What is the contractor's optimum decision policy for the problem illustrated in Figure 2.4 and explained in Example 2.4?

P7.2. In Problem P7.1 above the contractor hires an expert meteorologist whose ability to predict accurately normal water level, high water, and floods has the same conditional probabilities as given numerically in Table 7.1 for no, minor, and major hurricanes. For this situation determine the maximum fee the contractor would be willing to pay the meteorologist.

P7.3. A county highway engineer is allotted a fixed annual budget for maintenance of the county road network. The county road network contains different types of road surfaces and conditions.

The engineer is uncertain whether to plan the maintenance activities on a one-, two-, or three-year time horizon. He can postpone any activity for any road and tolerate deterioration in quality until the road must be replaced, initiate temporary or semi-permanent maintenance that either ensures the continuation of existing conditions or delays the time when the road must be replaced, or initiate a road replacement program with the same or better type roadways.

How would you go about formulating the problem as a n-year decision problem and what type of information would be required?

P7.4. A plant operator is about to commence an urgent above normal production run. To assure meeting the desired production rate, he has decided to transfer and incorporate into the production line a piece of equipment from another line at a cost of $5,000. He is considering whether to overhaul this equipment before placing it in the new production line.

The piece of equipment costs $800 to overhaul whereas if he incorporates the item into the process line and it then breaks down it will cost $1,500 to cover the cost of repair and lost time. He estimates that there is a 66 percent chance that the equipment motor is reliable but is assured that it will be reliable if it is overhauled. A dynamometer test of the motor costs $100 but will only indicate whether the motor is in good or bad condition with a 10 percent chance that the test results prove invalid. He estimates that there is a 70 percent chance that the dynamometer test will indicate a reliable motor.

1. Model the decision problem faced by the plant operator.
2. If the plant operator wishes to base his decision on an EMV policy, what should be his optimum strategy?

P7.5. A group of investors is considering two alternative plans for a high-rise building complex. Plan I calls for a 70-story apartment building and a separate adjacent 40-story office building. Plan II calls for a single 100-story high-rise building with 43 stories for offices and 55 stories for offices, stores, and shops and 2 stories for mechanical and operational systems. The estimated costs and lifetime return of the two plans are as follows:

PLAN I

ESTIMATED COST $(\times 10^6)$	PROBABILITY	ESTIMATED LIFETIME RETURN $(\times 10^6)$	PROBABILITY
$100	0.6	$300	0.5
95	0.3	250	0.4
90	0.1	200	0.1

PLAN II

ESTIMATED COST $(\times 10^6)$	PROBABILITY	ESTIMATED LIFETIME RETURN $(\times 10^6)$	PROBABILITY
$150	0.7	$450	0.2
120	0.2	350	0.4
100	0.1	250	0.3
		200	0.1

If the investors prefer Plan II, provided the expected net profit differs by less than $10 million, which plan should they choose?

P7.6. The Board of Supervisors for a county is planning to build a dam costing $5 million on Mountain Creek. To protect the dam, a separate spillway is required and the Board must decide whether to build a large spillway costing $3 million or a smaller spillway costing $2 million. Based on historical records, it is estimated that there is a 0.25 probability that one or more serious floods would occur during the life of the dam and a 0.10 probability that one or more major floods would occur. The probabilities that the two possible spillway types will fail during these two levels of floodings are estimated as follows:

	SERIOUS FLOOD		MAJOR FLOOD	
	Fail	Safe	Fail	Safe
Large spillway	0.05	0.95	0.1	0.9
Small spillway	0.10	0.90	0.25	0.75

If the spillway fails to function properly during a serious or major flood the dam will be destroyed. The replacement cost of the dam will be the same as the original cost. In addition to a total loss of the dam and its spillway, other property damages will be incurred. It is estimated that in case of failure during a serious flood, there is a 70 percent and 30 percent chance for other property losses amounting to $1 million and $3 million, respectively. In case of failure during a major flood, there is a 70 percent and 30 percent chance for other losses amounting to $3 million and $5 million, respectively.

1. Model this decision problem with a decision tree.
2. What is the optimum decision based on the EMV criterion?
3. How would you caution the Board about basing its decision on the EMV criteria?

Suppose now that the risk behavior of the Board is as follows:

a. Against a 90 percent chance of losing $7 million and a 10 percent chance of losing $20 million, the Board is willing to pay insurance of $12 million.

b. Against a 70 percent chance of losing $12 million and a 30 percent chance of losing $20 million, the Board is willing to pay insurance of $15 million.

c. Against a 50 percent chance of losing $12 million and a 50 percent chance of losing $20 million, the Board is willing to pay $17.5 million in insurance.

4. Draw a utility curve assuming a utility value of 50 for $-\$7$ million and a utility value of -100 for $-\$20$ million.

5. What should be the Board's decision in (2) when this utility curve is used?

P7.7. A heavy machinery manufacturing company is considering submitting a bid proposal for the supply and installation of the machinery system for the swing span of a bridge. The chief engineer of the company has estimated that a bid proposal can be prepared at a cost of $5,000, but that such a bid will have only a 20 percent chance of being accepted. As an alternative, the company can invest $20,000 on an extensive research study of the project before preparing the bid proposal. Such a proposal will have a 30 percent chance of being accepted.

If the company does get the job, it can either expand its personnel and staff or subcontract work to other companies. If a major portion of the work is to be performed by subcontractors, it is estimated that there is a probability of 70 percent, 20 percent, and 10 percent for making a profit of $1 million, $1.5 million, and $2.0 million, respectively. If the work is to be accomplished mostly by expanding staff, there is a probability of 60 percent, 35 percent, and 5 percent for profits of $0.5 million, $2 million, and $3 million, respectively.

1. Suppose that you were the company president and that your company has total assets of $10 million. Prepare a utility curve to reflect *your* risk characteristics.

2. Based on the above utility curve, what is your optimum policy?

3. What is the maximum dollar investment that you would be willing to provide for the preliminary proposed study?

P7.8. Discuss the subjective nature of the decision-making approach implicit in decision tree analysis? What alternative approaches exist for the decision-maker?

8

SYSTEM SIMULATION

Portion of the San Francisco Bay-Delta Model in Sausalito, one of the largest working hydraulic models of any harbor in the world. The one-acre iconic model has become a focal point for both environmental and engineering planning throughout the region since its completion in 1971. (Courtesy of the Civil Engineering Magazine, American Society of Civil Engineers)

8.1 SIMULATION CONCEPTS

The role of models in systems analysis is illustrated in Figure 8.1. Because engineering systems are so complex, a system model is first constructed to represent the real system and its environment. The model should include all the relevant system components and clearly define the component interrelationships, as well as indicate the constraints within and those imposed on the system. The designers have complete control over the components, structure, and constraints within their model. It is with this model that they test their designs and study the behavior of the system under various conditions.

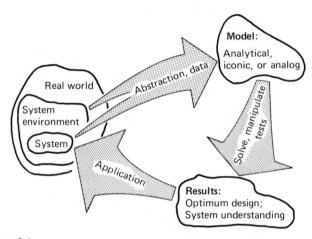

Figure 8.1

Role of Models in Systems Analysis

Simulation is the process of conducting experiments with a model of the system that is being studied or designed. It is a powerful technique for both analyzing and synthesizing engineering systems. In an analysis problem, the system model is generally fixed and the objective is to determine the system response when a set of input variables is allowed to take on different values. The simulation process is then basically an iterative procedure and may be described as an input-output study with feedbacks provided to guide the changes in the input parameters, as illustrated in Figure 8.2. The inputs define the set of events and conditions to which the system can be subjected in the real world, and the outputs predict the system response. By studying the outputs at the end of each iteration, the designer learns more and more about the system

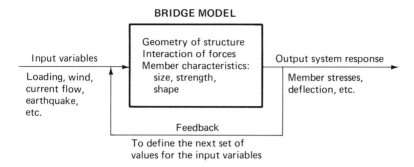

Figure 8.2

Analysis by Simulation

and may then use his newly acquired knowledge to define new sets of inputs to be processed through the model.

As an example, consider the problem of determining the failure conditions of a steel bridge. The simulation model may be in the form of a set of mathematical equations relating the interaction of the forces among the members, the geometric structure of the bridge, the size and tensile strength of the members, and so on; the input variables may include loading conditions, velocity and direction of wind, the flow velocity of water and debris in the channel, and the occurrence of earthquakes. For each specific set of combination of the input parameters, the model may be used to determine the stresses in the members and the deflection of the structure, which provide a direct measure of the performance of the system under the given set of conditions.

In system synthesis, the designer is interested in determining how the system components can best be put together so that the system can meet the performance standard. In this case, the system model itself is a variable; but a set of input-output characteristics has been specified as a design standard. The simulation process is again an iterative procedure, but the output of a simulation study is now used to decide which system parameter can best be changed in order to improve the performance of the system. As shown in Figure 8.3, the feedback is now directed to changes within the model itself. For example, in the above bridge example, the problem may be to choose an optimum combination of sizes and strengths for the structural steel members. These design parameters directly affect the strength of the total structure, its weight, as well as its cost. If one design fails to meet the performance standard according to one simulation result, these system parameters may be changed and the simulation repeated until a set of feasible alternatives have been established and an optimum solution is identified.

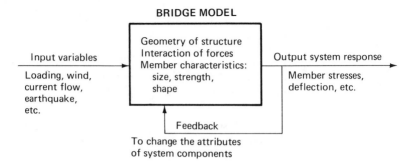

Figure 8.3

Synthesis by Simulation

The major advantage of the simulation approach lies in the fact that once a simulation model has been constructed, it may be used for both system analysis and synthesis and to test the design under a wide spectrum of environmental conditions. Moreover, after the design has been completed and the system implemented in the real world, the simulation model may be used to locate the sources of unpredicted system problems and to plan for system improvement.

8.2 SIMULATION MODELS

Simulation models can take many forms and can be of many different levels of complexity. A good model should represent the characteristics of the system so that the problem under consideration can be solved. Simulation models can be broadly grouped into three types:

1. Iconic;
2. Analog; and
3. Analytical.

Iconic models are physical replicas of the real systems on a reduced scale. This type of model is common in engineering. In aircraft design, wind tunnels are used to simulate the environment around an aircraft in flight. By subjecting a model of the aircraft to a well-chosen range of aerodynamic conditions, the designer can gain insight into the performance characteristics of the design. In the design of large engineering structures, such as skyscrapers, dams, bridges, and airports, three-dimensional architectural models are often prepared to provide a realistic view of the design. Such models are extremely useful both as a design tool and as a visual aid in presenting the project to the interested public. At the University of Illinois, for example, an indoor Watershed Experimentation System (Chow, 1968; Chow and Yen, 1968) is used to simulate the hydraulics of a watershed. By means of electronic computer control, the system is capable of producing rain storms of various time and space distributions moving in any desired direction. Using such a simulation model, the effect of a moving storm on the runoff distribution over any watershed can be experimentally studied.

Iconic models are particularly important for studying systems in which the interrelationship of the components are not well understood or in which components are too complex to be modeled mathematically. Thus, by constructing a replica of the system and experimenting with it under a set of controlled conditions, an insight can be gained into the system behavior. The end results may well be the discovery of some natural law that governs the components of that system.

In many engineering problems it is impossible to build a physical replica of the real system. For example, in studying the response of engineering structures to various intensities of earthquakes, it is impossible to build a small model of the earthquake zone using rocks and soils and to generate earthquakes at the command of the experimenter. However, if the dynamic property of quake waves is known, an instrument may be constructed to generate a similar type of force motion. At the University of Illinois at Urbana-Champaign, an electro-hydraulic system is used to simulate earthquakes (Sozen et al., 1969; Takeda et al., 1970). It consists of a 12-foot square table driven by a hydraulic ram and is designed to activate a 10,000 lb mass in one horizontal direction with a sinusoidal or random vibration. The simulation is used to determine the response of reinforced concrete structures to earthquake shocks. Such a simulation model, in which the real system is modeled through a completely different physical media, is called an analog model.

In problems in which the characteristics of the system components and system structure can be mathematically defined, an analytical model constitutes a powerful simulation tool. It may be composed of systems of equations, boundary constraints, and heuristic rules, as well as numerical data. Basically, the model consists of a set of design variables and a set of system constants. Some of the design variables are independent parameters, the values of which are specified as input to the simulation process. The remaining variables are dependent variables that are used to measure system performance and response and thus they constitute the outputs. However, to construct an analytical model requires that fundamental properties of the system components and their interactions be understood.

With the availability of high-speed, large memory electronic computers, analytical models are becoming versatile design aids in all disciplines of engineering. For example, analytical simulation models are playing an increasingly important role in air pollution control (Middleton, 1971). Although many electronic sensors have been developed to provide measurement of pollution levels, a sample network within a city is usually widely scattered and is not sufficiently dense to identify all heavily polluted areas. Moreover, such a system cannot predict the effect of new sources of pollution, such as proposed industrial plants or highways, on the overall community. Analytical simulation models have been a useful tool in both monitoring and predicting pollution levels. The model may consist of mathematical functions relating such system factors as the rates of pollutant decay, weather condition, topography of the locality, and the distribution of commercial, industrial, and residential districts. Once the major sources of pollutants are identified, such a model may be used to determine the pollution level anywhere within the community during different parts of the day and under different weather conditions. The simulation results may even include a contour map of the pollution level throughout the municipality. Moreover, once the model is constructed, it may also be used to study the effectiveness of the various methods of pollution control as well as the adverse effects of new pollution sources.

8.3 SIMULATING A CONCRETE PLANT OPERATION

To illustrate the method of simulation in general and that of analytical simulation in particular, consider a problem involving the operations of a company producing ready-mixed concrete. The company now owns a batching plant to mix the appropriate quantities of cement, sand, gravel, water, and special additives to produce concrete mixes that are then loaded into delivery trucks for delivery to the customer. The company now owns five delivery trucks. The company management has recognized that the full production capacity of the batching plant has not been utilized because of the limited number of trucks. Their problem is to determine the optimum number of delivery trucks that can be added to their fleet so that its batching plant can be fully utilized.

Figure 8.4 schematically illustrates the production process in the company's system. Company policy has dictated that orders be filled on a first-come-first-served basis, and in order to have all trucks back in time for clean up, no order is to be processed after 3:30 P.M. When an order is received at the central office, it is immediately placed in the waiting list of orders. Eventually, when its turn arrives for processing, the order is transmitted to the batching plant, which then initiates the following procedure. The batching plant operator calls the parking lot

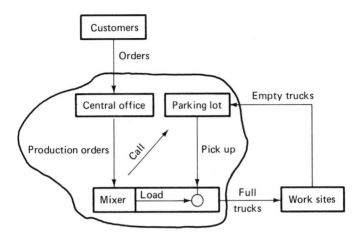

Figure 8.4

Ready-mix Concrete Plant Production Process

and requests the required number of trucks to fulfill the order. If delivery trucks are available, they are dispatched to the batching plant where the order is then loaded for delivery. However, if the required number of trucks are not available, those trucks that are available are loaded and dispatched to the delivery site. Then the batching plant has to stop operations and wait for trucks to return before completing the order. As a result of the existing shortage of trucks, there is a continuous backlog of orders and the batching plant is idle a large part of the day.

The general procedure of a simulation study of this problem is illustrated in Figure 8.5. A simulation model is constructed to represent

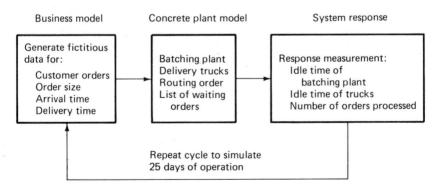

Figure 8.5

Simulation of a Concrete Plant Operation

the operational procedure within the concrete plant. A set of fictitious customer orders is generated to approximate the inflow of orders in the real system during each day. This set of fictitious orders is processed through the system, and the system response is measured according to the following set of parameters: idle time for the batching plant and the delivery trucks, number of orders processed, and number of truckloads delivered. Thus, by repeating the experiments for an extended simulation period of, say, six months, an accurate measure of the average idle time for the batching plant in one day may be obtained. Moreover, the same set of experiments can be repeated with a different number of trucks in the model. The outcome of the simulation process then provides a measure of the system response depending on the number of delivery trucks available.

Modeling Customer Orders

The purpose of modeling customer orders is to provide a means whereby a fictitious flow of customer orders can be generated for each day in the simulation process. The model should be representative of the real situation so that the flow of orders generated by it would closely resemble that normally encountered in the concrete plant. Within the scope of this problem, three elements of the customer orders are of major importance; namely, time between orders, number of truckloads in each order, and the round-trip delivery time required for the orders. Therefore, these elements should be included in each order generated by the model.

The data used in model construction can be obtained by direct sampling from the real system. The arrival times of all orders actually received during five successive days at the concrete plant are listed in column 3 of Table 8.1. In column 5, the sizes of the orders are listed in terms of the number of truck loads ordered. The time intervals between the arrival of two successive orders are computed and recorded in column 4. In columns 6 to 10, the actual time required for the delivery trucks to make a round trip are also recorded. The delivery time depends on the distance to the customer's construction sites, as well as on traffic conditions along the way. The latter factor accounts for the differences in time among trucks delivering to the same customer. Thus, the delivery time can be conveniently divided into two components. The average delivery time for an order depends primarily on the delivery distance. The deviation of each delivery from the average is due to unpredictable traffic conditions.

The data in Table 8.1 can now be used to develop models that can be used to generate a fictitious flow of customer orders. The first step is to develop a frequency distribution for each attribute to be included in a customer order. The data in column 4 can be used to develop the frequency histogram of Figure 8.6 for the arrival intervals. This is

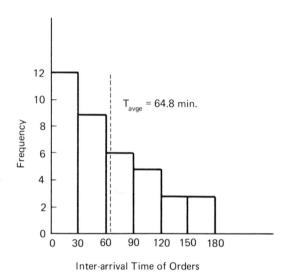

Figure 8.6

Histogram of Arrival Intervals

accomplished by grouping the data into classes. Thus, the data in column 4 has been grouped into six classes. The first class contains the 12 arrival intervals that are less than 30 minutes; the second class contains the arrival intervals that are between 30 and 60 minutes; the third class contains the arrival intervals between 60 and 90 minutes, and so on. Each class can be represented by its midpoint, e.g., each time interval in the first class can be represented by a time interval of 15 minutes. In Figure 8.7 the order size frequencies are converted to probability densities and shown on a pie chart. Figures 8.8 and 8.9 present the frequency histograms for average delivery time for an order and the deviation of each delivery from the average, respectively.

The models can be developed directly from these frequency distributions. For example, using Figure 8.6, thirty-eight pieces of paper of identical size can be prepared to represent the thirty-eight orders. On each piece the arrival interval before the next order is written. Thus,

Table 8.1

RECORDS FROM THE REAL SYSTEM

DAY	ORDER NO.	ARRIVAL TIME	INTERVAL FROM LAST ORDER	NUMBER OF TRUCK-LOADS	ROUNDTRIP DELIVERY TIME (MINUTES)						DEVIATION FROM AVERAGE
					Truck-load 1	Truck-load 2	Truck-load 3	Truck-load 4	Truck-load 5	Average	
Monday	1	8:15	15	2	64	69				67	$-3, 2$
	2	8:32	17	1	75					75	0
	3	9:03	31	3	73	77	70			73	$0, 4, -3$
	4	9:47	44	3	123	116	123			121	$2, -5, 2$
	5	10:00	13	2	105	104				105	$0, -1$
	6	11:16	76	4	27	34	34	30		31	$-4, 3, 3, -1$
	7	1:03	107	1	48					48	0
	8	2:06	63	2	55	59				57	$-2, 2$
	9	2:18	12	1	83					83	0
	10	2:49	31	3	50	46	54			49	$1, -3, 5$
	11	4:06	77	2	61	65				66	$-5, -1$
Tuesday	12	9:40	100	3	73	78	77			76	$-3, 2, 1$
	13	10:12	32	4	66	71	70	69		69	$-3, 2, 1, 0$
	14	10:45	33	4	41	43	45	48		44	$-3, -1, 1, 4$
	15	11:55	70	3	60	65	65			63	$-3, 2, 2$
	16	1:24	89	2	91	103				97	$-6, 6$
	17	2:22	58	1	7					7	0
	18	4:27	125	3	99	105	94			99	$0, 6, -5$

Table 8.1 (Continued)

DAY	ORDER NO.	ARRIVAL TIME	INTER-VAL FROM LAST ORDER	NUM-BER OF TRUCK-LOADS	ROUNDTRIP DELIVERY TIME (MINUTES)						DEVIATION FROM AVERAGE
					Truck-load 1	Truck-load 2	Truck-load 3	Truck-load 4	Truck-load 5	Average	
Wednesday	19	8:25	25	4	47	41	40	45		43	4, −2, −3, 2
	20	10:35	130	4	68	66	62	66		66	2, 0, −4, 0
	21	12:03	88	3	88	88	85			87	1, 1, −2
	22	3:02	179	5	53	59	52	53	53	54	−1, 5, −2, −1, −1
	23	3:50	48	2	59	38				39	0, −1
	24	4:15	25	5	30	39	32	37	32	34	−4, 5, −2, 3, −2
Thursday	25	8:05	5	4	74	71	70	72		72	2, −1, −2, 0
	26	10:33	148	3	63	62	63			63	0, −1, 0
	27	1:04	151	2	94	88				91	3, −3
	28	1:28	14	4	81	87	88	82		85	−4, 2, 3, −3
	29	1:40	12	3	38	40	41			40	−2, 0, 1
	30	2:23	43	1	21					21	0
	31	4:10	107	4	71	69	69	70		70	1, −1, −1, 0
Friday	32	8:30	30	5	72	76	75	73	69	73	−1, 3, 2, 0, −4
	33	9:28	38	4	107	94	100	94		99	8, −5, 1, −5
	34	12:12	164	4	60	69	66	65		65	−5, 4, 1, 0
	35	12:33	21	3	55	57	54			55	0, 2, −1
	36	2:05	92	1	82					82	0
	37	2:15	10	4	87	91	85	82		86	1, 5, −1, −4
	38	4:13	118	2	55	59				57	2, −2

Number of truck loads (x)	Number of orders	Probability p(x)
1	6	0.158
2	8	0.211
3	10	0.263
4	11	0.289
5	3	0.079
	Total = 38	

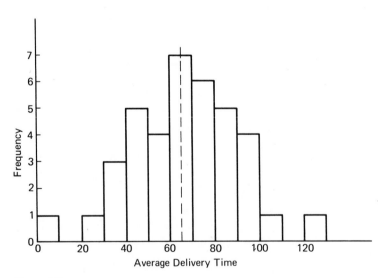

p(x) = Number of orders for x truck loads/38

Figure 8.7

Model of Order Sizes

there should be twelve with the number 15, nine with the number 45, and so on. These paper pieces are then placed in a cup. This cup is now essentially a simulation model for generating arrival intervals for customer orders.

A similar model can be constructed for the distribution of order sizes. According to Figure 8.7, the cup should contain six paper pieces

Figure 8.8

Histogram of Truck Delivery Times

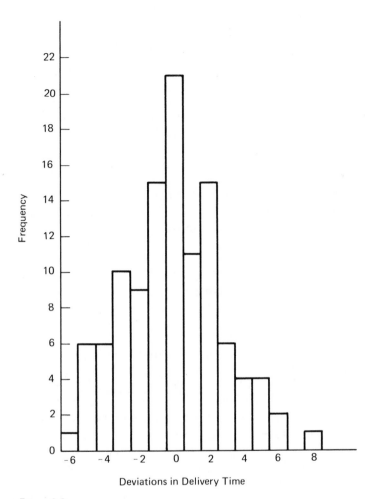

Figure 8.9

Histogram of Deviations in Delivery Time

marked with 1 truckload, eight marked with 2 truckloads, ten marked with 3 truckloads, and so on.

A third cup should contain thirty-eight paper pieces representing the thirty-eight average delivery times according to the frequency distribution in Figure 8.8. This cup of paper pieces is then the simulation model for generating the average delivery time of an order. Similarly, a fourth cup should contain the 111 pieces of paper representing the deviations from average according to Figure 8.9.

To generate one fictitious customer order, the cups are covered and shaken thoroughly so that the paper pieces are well mixed. One piece is then drawn from each of the first three cups. The piece from the first cup gives the arrival interval of the order; the second gives the number of truckloads for the order; the third gives the average delivery time. These paper pieces are then returned to their respective cups. To generate the deviation in delivery time for each truckload, a paper piece is drawn from the fourth cup. The time marked on it is recorded and then returned to the cup before a second paper piece is drawn for the next truckload. Figure 8.10 gives a listing of jobs generated for one day of simulated operation.

The above procedure can be repeated as many times as necessary. The reliability of the simulated data depends on how representative the original sample data are of the real situation. In order to improve the reliability of the models, it is necessary to collect actual sample data from at least several days of normal operation.

Modeling the Plant Operations

The simulation model of the operations within the concrete plant will consist of a series of logistic rules that describe the routing and processing of customer orders through the plant. Figure 8.11 is a general schematic simulation model of the plant operations. In addition, the model should provide a mechanism whereby the operational status of the batching plant and delivery trucks can be continuously recorded. For example, the time diagram in Figure 8.12 may be used to supplement the schematic model in Figure 8.11. Also shown in Figure 8.12 are the results for one day of simulated operation using the list of orders generated in Figure 8.10. During the day, the batching plant is idle 58 percent of the working time and the trucks idle on the average of 28.4 percent of the working time.

The above results represent one simulated day. In order to increase the reliability of the results, many days must be simulated. The mean of all the results is then used to indicate the operational characteristics of the system when only 5 delivery trucks are available.

Simulation can be extremely tedious when computations are performed manually, as in the above example. However, the schematic model in Figure 8.11 can be easily programed for computer use, as described in Section 8.4. Using a modern, high-speed electronic computer, one day of simulated operation can be performed in much less than

Order Number	1	2	3	4	5	6	7	8	9	10	11	12
Arrival interval (min.)	15	15	45	15	75	15	165	135	15	15	15	15
Arrival Time	8:15	8:30	9:15	9:30	10:45	11:00	1:45	4:00	4:15	4:30	4:45	5:00
No. truck loads	3	2	5	3	1	4	2	4	3	5	4	2
Average delivery time	45	105	85	95	75	65	65	125	85	45	105	95
Deviation from average delivery time for each truck — 1	2	-5	5	1	3	4	-2	-5	-1	2	2	0
2	0	1	0	-2		-3	5	1	-1	-3	2	-1
3	2		-1	-2		-5		2	2	-3	0	
4			-1			-2		-1		0		
5			1							5	0	

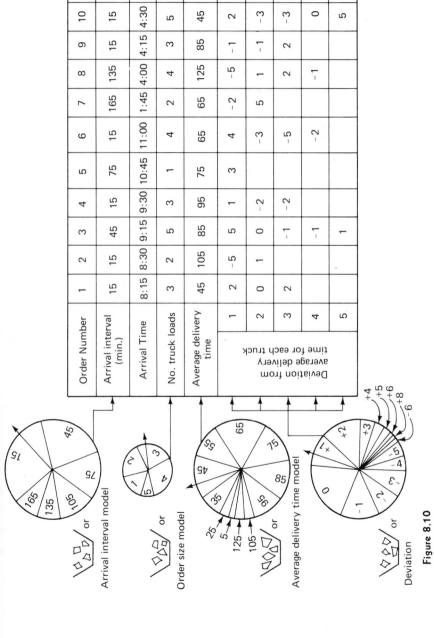

Figure 8.10

Simulated Customer Orders for One Day

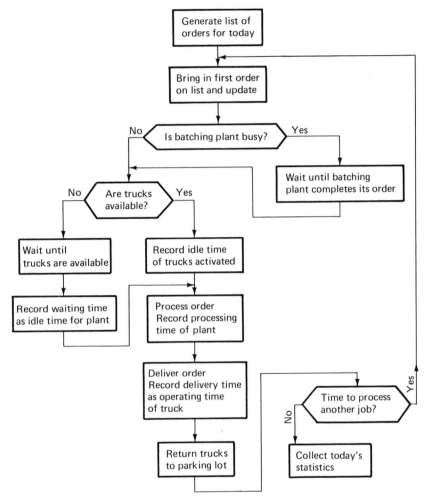

Figure 8.11

Flow Chart for Simulation Model of Concrete Plant Operations

1 second of computer time. Therefore, the power of analytical simulation relies almost exclusively on the availability of an electronic computer.

Another important lesson should be learned from the example in Figure 8.12. The first order of the day arrived at 8:15 A.M., and hence all equipment was idle during the first 15 min of working time. This is normally not the case, since there should always be some orders left over from the preceding day. To overcome this problem, a so-called *run-in*

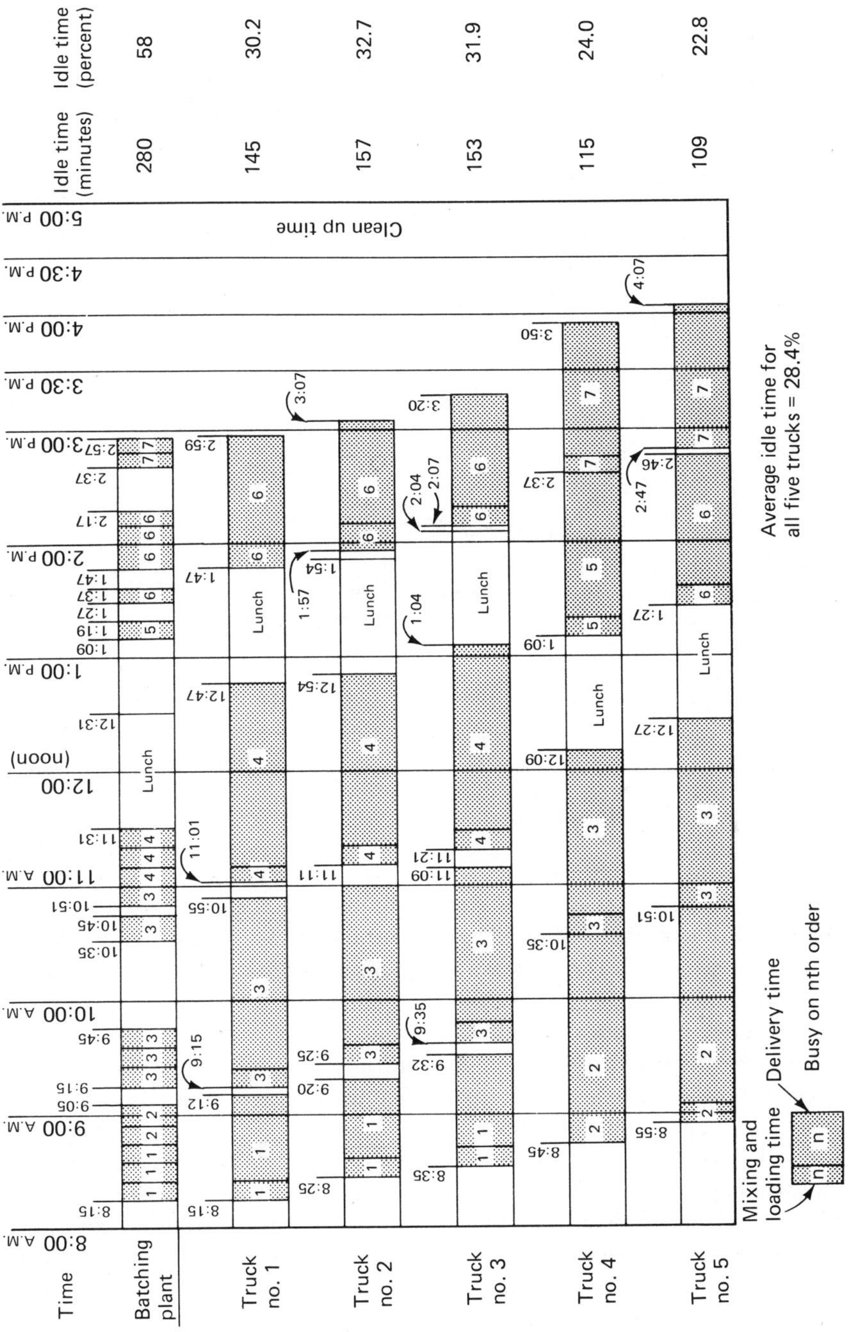

Figure 8.12

One Day of Simulated Operation

period of several simulated days should be allowed before statistics are recorded on the system response characteristics.

The Simulation and Analysis Process

Once the simulation model has been constructed, it is then a simple matter to use as many delivery trucks in the model as desired. Figure 8.13 shows a possible result of this simulation problem. Each

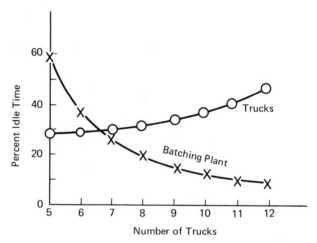

Figure 8.13

Analysis of Simulation Results

point in the graph is the average computed from a large number (say 100 days) of simulated days. The percentage of idle time of the batching plant decreases with the increase in the number of delivery trucks. However, beyond a certain point, the rate of decrease becomes very small. Furthermore, because of operational constraints, such as the rule that no orders will be processed after 3:30 in the afternoon unless the delivery truck can return to the plant by 4:30 p.m., the percentage of idle time never approaches zero.

Contrary to the response of the batching plant, the idle time of the delivery trucks increases with the increase in the number of trucks. Therefore, the efficiency of one component of the system is sacrificed for the efficiency of another system component. Undoubtedly, the management will want to consider also the economics of the problem, since the ultimate objective is to maximize the profit for each dollar invested. The computation of capital and operational cost as well as daily sales volume for different numbers of delivery trucks can be easily incorporated into the above simulation model. Having determined all the relevant system response characteristics from simulation, the company management must use their results to find an optimum trade-off for the problem. The optimization techniques, such as mathematical programming and decision analysis, can be used to perform this function.

8.4   GENERATING INPUT DATA

A simulation study requires a sequence of generated input data with the same frequency distribution as the sequence of sample input data. In the concrete plant example, this was accomplished by drawing pieces of paper from a cup. Each piece of paper had a number written on it and the frequency distribution of the numbers in each cup was the same as the frequency histogram for the input attribute that was being simulated. Roulette wheels, random number tables, or numerical methods can also be used to generate input data for simulation.

Roulette Wheels

Consider, for example, the frequencies of order sizes in Figure 8.7. The frequencies can be transformed into probability densities by determining the relative probability of the occurrence of each class; e.g., class 1 has a probability of occurrence $\frac{6}{38}$, or 0.158. A roulette wheel can now be constructed such that the divisions on the wheel are proportional to the probability of each class, as shown in the pie chart in Figure 8.7. Now, in order to generate an order size, we just spin the wheel and record the number on which it stops. The same can be done for each of the other frequency histograms.

Random Number Tables

Random number tables are groups of ten numbers (0 to 9), or hundreds of numbers (00 to 99), or any other larger similar group. The order in which the numbers are placed is absolutely random. Appendix B is a table of random numbers between 000000 and 999999. These numbers may be thought of as the result of spinning a roulette wheel that is marked with 1,000,000 equal divisions and labeled 000000 to 999999. Each number is equally likely and appears at random.

In order to use a random number table to generate input data, the random numbers must be assigned to each class of the input data in proportion to the probability of occurrence of that class. Consider again the probability model of Figure 8.7 for order sizes. The first class has a probability of occurrence of 0.158 and, therefore, 15.8 percent of the random numbers should be assigned to the class representing one truckload. This can be accomplished by assigning the numbers 000000 through 157999 to this class. The next class (2 truckloads) has a probability of occurrence of 0.210 and, therefore, the numbers 158000 through 367999 should be assigned to this class. Similarly, the third class (3 truckloads) is assigned 368000 through 630999, the fourth class (4 truckloads) is assigned 631000 through 920999, and the fifth class (5 truckloads) is assigned 921000 through 999999.

We can now draw sampling numbers from the random number table (Appendix B). From the first column, we find the sequence 258164, 547250, 279794, and so on. Each of these numbers represents an order size. Thus, 258164 corresponds to an order size of 2 truckloads. Similarly, 547250 and 279794 represent order sizes of 3 and 2 truckloads, respectively. We began at the first number in the table and began to read down the column, but we could have begun anywhere in the table. Furthermore, it is not necessary always to read down the column. As long as the numbers are picked from the table in some regular fashion, such as down a column or from a horizontal row, they will appear at random with no member in the group enjoying any preference over any other.

The probability distribution of the order sizes will be the same in the generated sequence as in the sample sequence because each order size has random variables assigned to it in proportion to its probability of occurrence in the sample.

This same procedure can be used to generate the input data for arrival interval times, average truck delivery times, and deviations from average delivery times.

Numerous tables of random numbers are available, of which the most extensive are those published by the RAND Corporation (1955).

Numerical Methods

When simulation is performed with the help of electronic computers, it is convenient to describe the frequency histograms by appropriate mathematical functions. For example, the frequency histogram for the inter-arrival time of orders in Figure 8.6 may be adequately described by the exponential density function

$$F(t) = \begin{cases} \dfrac{1}{t_a} e^{-\left(\frac{t}{t_a}\right)} & \text{for } t > 0 \\ 0 & \text{elsewhere} \end{cases} \tag{8.1}$$

where t is the time interval between successive orders, and t_a is the average time interval, which is equal to 64.8 minutes in this example. Similarly, the histogram in Figure 8.8 can be described by a normal density function as follows:

$$f(D) = \frac{1}{\sqrt{2\pi}\sigma_D} e^{-\left(\frac{D-u}{\sigma_D}\right)^2} \tag{8.2}$$

where D is the average delivery time; u is the mean value of D and is equal to 66.1 minutes in this example; and σ_D is a constant that defines the lower and upper limit within which 68 percent of the values of D lie. Similarly, the following normal density function may be used to describe the histogram in Figure 8.9:

$$f(d) = \frac{1}{\sqrt{2\pi}\,\sigma_d} e^{-\left(\frac{d}{\sigma_d}\right)^2}$$

where d is the deviation in delivery time.

Since we can seldom afford to fill the computer's memory with a large number of random digits, computer programs are available to generate numbers that conform to these and other distribution functions (Balintfy et al., 1966; IBM, C20-8011).

8.5 THE SIMULATION PROCESS

The previous example of a concrete plant operation demonstrated the general procedure of system stimulation. The major procedure of system simulation study may be summarized as follows:

Problem Definition

A simulation study is usually conducted to serve two primary objectives:

1. To measure the system response under a wide range of system inputs;

2. To measure the system response when the system components and their interrelationship themselves are altered.

Therefore, in defining the problem it is vital to clearly identify the input parameters, the system components, their interrelationship, and the feasible alternatives of system design to be investigated. In addition, a set of quantifiable parameters, which can realistically characterize the system response, must be specified.

In the above example, the input parameters are the customer orders and their elements, which include arrival interval, number of truckloads in each order, and the round-trip delivery time. The system components include the batching plant and the delivery trucks, and the interrelationship of the components is established by the routing procedure. Finally, the parameters selected to measure the system response are idle time of the batching plant and the delivery trucks.

System Input Model Construction

Models must be constructed to simulate the behavior and pattern of the input parameters in the real system. The purpose of these models is to generate fictitious input data in the simulation process. The models may be composed of mathematical functions, logical steps, or numerical data. They are usually constructed from a set of sample data collected from the real system.

System Model Construction

A model is then constructed to realistically represent the operation of the real system.

Solution Process

The solution process comprises two major cycling steps, as shown in Figure 8.14. In the inner cycle, the system response is measured with respect to a specific set of values for the design parameters within the system model. In order to test the design rigorously under the complete feasible range of the input parameters, a large number of independent

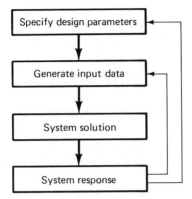

Figure 8.14

The Solution Process in Simulation

solutions must be conducted within this cycle. In the concrete plant example, each inner cycle represents only one simulated day, which may or may not be representative of the system response. In order to provide a reliable measure of the response trend over an extended period of operation, many days must be simulated. Alternatively, management may wish to determine the system response to an increase in the flow of customer orders. This constitutes an inner loop cycling process and the results will indicate the sensitivity of the system to changes in input data.

In the outer cycle, the system response is measured with respect to changes in the design parameters. If the simulation model is properly constructed, the values of the design parameters can be easily changed within the model. In the concrete plant example, the only variable design parameter is the number of delivery trucks. Thus, many days are first simulated using five trucks. The entire procedure is then repeated with six trucks, seven trucks, and so on. Studies of this nature indicate the sensitivity of the system to changes in the values for system parameters.

In some simulation problems, it is necessary to study the system response using several different system structures. It may then be necessary to build separate system models to represent the different designs. For example, if the management of the concrete plant wishes to study the relative benefits of several different job routing procedures in addition to the first-come-first-served method, it will be necessary to build a separate system model for each routing method. In these cases, the structure and logic of the system will vary. Studies of this nature indicate the sensitivity of the system to changes in system structure.

Analyzing Simulation Results

Simulation is neither an optimization nor a decision-making process. It merely provides an understanding of the system response under a set of specified conditions. The results must then be carefully analyzed using system techniques to establish an optimum design or policy.

8.6 LIMITATIONS OF THE SIMULATION APPROACH

The simulation approach provides a powerful tool for analyzing complex systems that cannot be easily studied by any other means. However, the accuracy and reliability of the simulation results depend largely on how well the simulation models can truly represent the response characteristics of the system in the real world. On the other hand, the validity of the models and the reliability of the results are extremely difficult to evaluate because of the very complexity of the system. It is advisable, therefore, that the simulation results be always analyzed with a certain degree of reservation, and that the response of the system in the real world be continuously monitored so that the feedback from the real world may be used to update and validate the simulation models.

Because of the above limitations, the simulation approach should, wherever possible, be supported concurrently by a theoretical study of the system; the latter serves to provide a hypothesis or simplified mathematical model of the system response characteristics. The results from the two approaches then serve as mutual checks. Moreover, even a simple theoretical analysis can help to narrow the values of the input and design parameters to be tested in the simulation process and thus effectively reduce the simulation calculations.

8.7 PROBLEMS

P8.1. Simulate two days of operations for the concrete plant example described in this chapter. Generate the necessary data using the random number tables of Appendix B as explained in Section 8.4.

P8.2. By direct observation determine the probability of inter-arrival times for vehicles at an intersection. For what period of time is your model valid, and what factors might affect the probability distribution you have obtained?

P8.3. Figure P8.3 illustrates an entrance ramp that leads into a four-lane interstate highway. The frequencies of arrivals at the ramp-highway intersection during peak hours are as follows:

EAST-BOUND TRAFFIC ALONG HIGHWAY		RAMP TRAFFIC	
Arrival Interval (seconds)	*Frequency (cars)*	*Arrival Interval (seconds)*	*Frequency (cars)*
0	0	0	0
2	50	4	2
4	40	8	2
6	30	12	4
8	30	16	3
10	25	20	5
12	25	24	6
14	15	28	4
16	10	32	6
18	10	36	8
20	5	40	10

There is a 0.75 probability that a car arriving in an east-bound lane will be in lane 1. To enter safely, a vehicle waiting to enter the highway requires 10 seconds before the next car in the east-bound lane 1 arrives.

1. Simulate one 30-minute period of traffic flow to determine the average waiting time for the vehicles entering from the ramp.

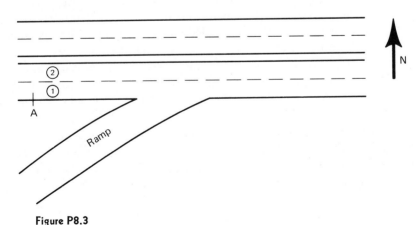

Figure P8.3

Also, determine the maximum number of cars waiting at any one time.

2. The traffic department is considering the possibility of installing a traffic light at point A that will direct cars into lane 2 such that the probability of a given car being in lane 1 becomes 0.25. Determine the ramp waiting time and maximum queue length with the light installed.

P8.4. The development of a new plant is under consideration and the probability distributions for order inter-arrival times and order sizes are as follows:

ORDER INTERVAL (DAYS)	PROBABILITY	ORDER SIZE	PROBABILITY
0	0	10	0.05
0–1	0.05	50	0.20
1–2	0.15	100	0.05
2–3	0.35	200	0.20
3–4	0.20	300	0.40
4–5	0.15	400	0.05
5+	0.10	500	0.05

Management is concerned with the problems associated with the selection of production rate and finished product inventory size.

1. For a given production rate of 350 units per 5-day week and inventory size of 300 units, determine how frequently the orders cannot be immediately fulfilled and the average delay. Simulate the plant operation for a period of 2 months.

2. For the same production rate, what would be the maximum inventory size to ensure that all orders can be immediately filled on receipt? Simulate the plant operation for a period of 2 months.

P8.5. A building owner has decided that a single elevator in his building is not adequate and causes unnecessary delay for people going to offices in his building. You are asked to study the problem and make recommendations. You plan to simulate the system behavior with an analytic model. The study can be divided into three steps:

1. Model the present system with one elevator,
2. Test this model of the present system to see if it is a reasonable representation of the real thing,
3. Study variations in the model to determine their effect.

In this problem you are to work out in detail step 1. Consider carefully what should be the input and output for the simulation model. Draw a flow diagram of your simulation model. The output should be in the form of numbers, and you are to use the simulation model to derive the output. Then discuss how you would proceed in step 2. For step 3 repeat the analysis with two elevators and the same input to see how the output from the simulation model is affected.

Tables P8.5a and P8.5b contain data that was collected at the entrance to the elevator on the first floor. You may assume that this data

Table P8.5a

NUMBER AND ARRIVAL TIME OF PASSENGERS
DURING OBSERVATION PERIOD

ARRIVAL TIME* (MINUTES)	NUMBER OF PASSENGERS	ARRIVAL TIME* (MINUTES)	NUMBER OF PASSENGERS
0		22.25	2
2.00	1	24.25	3
3.25	3	24.75	2
4.75	2	26.00	4
6.50	3	27.00	2
7.50	5	28.50	2
9.00	2	29.25	5
10.00	3	31.00	3
11.25	1	32.00	3
13.00	3	33.50	4
14.00	2	34.00	2
14.75	2	34.75	4
16.00	1	36.50	1
17.50	3	37.50	1
18.75	4	38.75	3
19.50	2	39.00	2
20.75	4		

* Time is started at zero for convenience; you may convert to clock time with an arbitrary starting time if you wish.

is typical of the traffic entering the building and for the performance of the elevator. Everyone entering the building would like to use the elevator. This is a very small elevator, however, with a maximum capacity of five people and notice that it is quite slow. These constraints are imposed to reduce the amount of calculations required but do not detract from the general principles involved.

Table P8.5b

FREQUENCY OF TRIP DURATIONS OF THE ELEVATOR WITH 1 TO 5 PASSENGERS

		$\frac{1}{4}$	$\frac{1}{2}$	$\frac{3}{4}$	1	$1\frac{1}{4}$	$1\frac{1}{2}$	$1\frac{3}{4}$	2	$2\frac{1}{4}$	$2\frac{1}{2}$	$2\frac{3}{4}$	3
		TRIP DURATION OF THE ELEVATOR (MINUTES)											
Number of Passengers	1	15	35	44	48	45	40	7					
	2			15	28	38	43	35	26	15	7	2	
	3				10	20	28	35	38	32	23	14	8
	4					4	8	14	22	30	34	31	25
	5						4	7	11	15	21	27	30

Table P8.5b (cont.)

		$3\frac{1}{4}$	$3\frac{1}{2}$	$3\frac{3}{4}$	4	$4\frac{1}{4}$	$4\frac{1}{2}$	$4\frac{3}{4}$	5	$5\frac{1}{4}$	$5\frac{1}{2}$	$5\frac{3}{4}$	6
		TRIP DURATION OF THE ELEVATOR (MINUTES)											
Number of Passengers	1												
	2												
	3	3											
	4	21	16	13	9	6	4	2					
	5	28	26	23	21	18	16	13	11	9	6	3	1

P8.6. Table P8.6 lists a 20-year record of flood damage for a river basin. In designing a flood control project for the area, the amount of capital investment to be spent on flood control is related to the anticipated reduction in flood damage as a result of the flood control project. Assume that the cost of flood damage can be related to the discharge rate by the following simple expression:

$$D = KQ$$

where D is the flood damage in dollars; Q is the river discharge rate in cu. ft./sec.; and K is a multiplication factor, the value of which depends on the time of the flood, concentration and distribution of residential and industrial areas, etc. Assuming that there is only one flood each year, simulate a 50-year period and compute the total flood damage for the 50 years. (Hint: First develop probability distributions for K and Q from the given data.)

Table P8.6

20-YEAR FLOOD DAMAGE RECORD

YEAR	DISCHARGE RATE (cfs)	DAMAGE ($ millions)	YEAR	DISCHARGE RATE (cfs)	DAMAGE ($ millions)
1951	1,000	10	1961	2,300	18.5
1952	2,500	30	1962	3,400	34
1953	4,200	25	1963	4,500	50
1954	5,700	40	1964	1,300	8
1955	1,200	12	1965	2,000	14
1956	3,600	28.5	1966	6,000	60
1957	5,000	70	1967	1,700	17
1958	2,600	26	1968	3,500	28
1959	7,000	70	1969	4,300	56
1960	1,500	195	1970	3,000	30

P8.7. Three stockpiles of crushed aggregate of fine, medium, and coarse grades receive rock from a crusher and deliver rock to customers' trucks.

The rock arrives ordinarily in the stockpiles at a uniform total rate of 600 tons per day over 6 hours commencing at 7 A.M. The crushed rock is delivered from the crusher normally in the proportions of 3:2:1 for fine, coarse, and medium, respectively.

When any stockpile drops to 50 tons or less a special order is placed to the crusher to provide an additional 50 tons of the grade required using spare crusher capacity only. These additional orders flow in at the rate of 50 tons per hour, commencing as soon as a request is made. If such an order is incomplete at the end of the working day, it is completed first thing the following working day.

If any stockpile reaches or exceeds 500 tons, action is taken as follows:

1. For the fine-grade stockpile 50 tons is disposed of into five 10-ton trucks as soon as possible from the same loading point that supplies customers' trucks. These five trucks are immedi-

ately available when needed and each requires 10 minutes to be loaded. However, they must always give priority to customers' trucks.

2. For the medium-grade stockpile, supply from the crusher is completely stopped until the stockpile is reduced to 450 tons. This stoppage would not apply to the crushing and delivery of a special 50-ton order to replenish a depleted stockpile.

3. For the coarse-grade stockpile the proportion of the three grades from the crusher is immediately altered from the normal 3:2:1 for fine, coarse, and medium, respectively, so that all rock goes into fine and medium grades in the proportions 4:2, respectively. However, when this alteration is made the total crusher rate of supply is reduced to 50 tons per hour, excluding special orders of 50 tons to replenish depleted stockpiles. This altered output is allowed to continue until the coarse stockpile is reduced to 400 tons or less.

The arrival of a truck represents an order and orders are supplied in full truckloads of one grade only. Trucks are either of 5 tons or 7 tons capacity, arriving in equal proportions. The mean rate of truck arrival is 100 trucks per 8-hour day starting at 8 A.M. The mean rate is uniform throughout the day and trucks arrive according to the following probability distribution:

INTER-ARRIVAL TIME (Minutes)	PROBABILITY
0–1	.02
1–2	.03
2–3	.10
3–4	.15
4–5	.25
5–6	.15
6–7	.10
7–8	.10
8–9	.05
9–10	.05
10+	.00

Only one truck can be loaded at a time at each stockpile, with a total service time of 10 minutes.

The demand for the grades of fine, medium, and coarse rock is in the proportions 2:3:1.

1. By means of a schematic diagram, show how this rock-crushing plant operation can be modeled for the purpose of simulation. The diagram should illustrate the following:
   a. Major system components,
   b. Component characteristics, and
   c. Component interactions.

P8.8. Simulation can be used to study the response of an existing system for different levels of input or it can be used to test alternative designs of proposed systems.

Discuss the following subjects with respect to each of the above simulation problems.

1. What level of detail is to be used and what variables should be considered?

2. How are the results to be used?

3. What confidence can be placed in results obtained using short observation periods for the characteristics of the selected variables and of the simulation runs?

9

SYSTEM PLANNING

The New York John F. Kennedy International Airport was selected as the Outstanding Civil Engineering Achievement of 1961. The complex covers an area of 4,900 acres. It is designed to handle nearly all the New York Metropolitan Area's international air traffic, half its domestic long-haul air traffic, and one fourth of its domestic short- and medium-haul air traffic. The terminal city, shown above, is the core of the airport and includes ten terminals. (Courtesy of the American Society of Civil Engineers)

9.1 SYSTEM FEASIBILITY

Well-established procedures exist for initiating and developing engineered systems. The owner (government, company, or individual) of the proposed system employs or contracts with a professional engineer to whom he explains his problem as he sees it, usually in terms of a proposed system. Then the engineer, in cooperation with the owner, should review the owner's goals and objectives to determine if the proposed system is compatible with these goals and objectives or whether a different system is really what the owner wants. After this step, the engineer evaluates the alternative designs applicable to the proposed system.

The goals and objectives will depend on the problem or problems of the owner and the evaluation procedures will depend on the type of system. However, each proposed system will usually be required to pass tests for engineering, economic, financial, political, and social feasibility.

Engineering feasibility requires that the proposed system be capable of performing its intended function. Design analysis procedures as described in standard engineering texts can be used to indicate the ability of a proposed system to perform its intended function. In addition, the construction or implementation of the system must be possible.

A proposed system is economically feasible if the total value of the benefits that result from the system exceed the costs that result from the system. Economic feasibility depends on engineering feasibility because a system must be capable of producing the required output in order to produce benefits.

The owner must have sufficient funds to pay for system installation and operation before the proposed system is considered to be financially feasible. Financial feasibility may or may not be related to economic feasibility. An owner may be able and willing to pay for a system in order to fulfill noneconomic goals. It may also be that an economically feasible project is financially infeasible because the owner is not able to raise enough money to implement the system.

Political and social feasibility is assured if the required political approval can be secured and if the potential users of the system will respond favorably to system construction. Every system is subject to review at different stages of planning. A private company has executive officers or a board of directors that reviews proposed systems. Public systems are subject to public hearings and review by committees of elected representatives. Usually political support is gained after evidence of engineering and economic feasibility has been presented. However, political pressure may be quite strong for a specific system even if it is economically infeasible. Conversely, groups that feel that they are adversely affected often oppose economically feasible systems because noneconomic factors have not received sufficient emphasis. Political and social feasibility can best be attained by active participation of representatives from all interested groups in planning and designing a proposed system.

9.2 PLANNING HORIZON

The planning horizon is the most distant future time considered in the engineering economic study. The planning horizon to be used for a specific study depends on the purpose of the planning and the scope or areal extent for which the planning is done.

If the purpose of the planning is to develop a framework plan that is to serve as a guideline for development, a time 50 years or more in the future may be chosen as the planning horizon. The framework plan will indicate the types of systems needed to achieve the desired results from the development. A framework plan for a river basin region might include the number of reservoirs and other types of water resources development, along with the purposes that each of these elements should serve in order to sustain the projected economic and population growth of the region for the next 50 years.

A planning horizon of 20 to 30 years in the future may be chosen for a study concerning investment decisions. The size and type of water treatment plant that a city will invest in may be based on existing water treatment methods and the projected economic and population growth of the city for the next 20 years. Many systems may have even shorter planning horizons. The construction of a large dam may require that a concrete mixing plant be constructed at the dam site to provide concrete for the dam. The size and type of mixing plant to be used may be determined from a study with a planning horizon that coincides with the projected completion date of the dam, some 1 to 5 years in the future. In some cases, the planning horizon for a major plant straddles several projects, in which case the salvage value of the plant at the end of a particular project may influence both its economic and financial feasibility.

The scope or areal extent of a proposed system will also be a factor in determining the planning horizon to be used in a study. The planning horizon for a rural farm-to-market road will usually be shorter than for an interstate highway system.

Traditionally, three different periods have been considered in planning studies: the physical life of the system, the economic life of the system, and the period of analysis. The physical life of a system ends when it can no longer physically perform its intended function. The physical life of a building does not end if the building is converted from a hotel to a museum. Its physical life ends when it can no longer provide shelter or support the loads sustained in the use of the building.

The economic life of a system ends when the incremental benefit from continuing operation of the system one more time period no longer exceeds the incremental costs of continuing operation one more time period. This point usually occurs when the annual operation, maintenance, and repair (OMR) costs equal or exceed the annual benefits from the system. Since a program of regular maintenance and periodic replacement of worn parts may extend the physical life of a system almost indefinitely, the economic life is usually shorter than the physical life.

The period of time over which the system consequences are considered to affect the system benefits and costs is referred to as the period of analysis. The uncertainty associated with future benefits and costs increases as the length of the period of analysis increases. In actual problem-solving, it is necessary to confine studies to a certain extension in space and time. This confinement within specific limitations is required because of the uncertainty associated with future benefits and costs and because of our inability to trace and evaluate the consequences of the system beyond the immediate vicinity of the system. Regardless of the period of analysis used, the salvage value of the system should be considered in the analysis. The salvage value is the worth of the system at the end of the period of analysis. For some systems, this worth will be positive, indicating that the system still has value or can be sold and thus can be thought of as a benefit that occurs at the end of the period of analysis. For other systems, this worth may be negative, indicating that it is a liability that must be disposed of and thus can be thought of as a cost that occurs at the end of the period of analysis.

When alternative schemes of development are being considered for the same purpose, all alternatives must be evaluated for the same period of analysis. If some component requires periodic replacement, then its cost is usually assumed to be repeated at the end of each component life until the total period of analysis is completed. However, this assumption should not be made without considering the effects of inflation, the development of new production techniques through technological advance, and the changing nature of demand for the system outputs with time. Uncertainty about any of these factors tends to favor alternatives with short lives.

9.3 ECONOMIC ANALYSIS

Costs and benefits associated with a proposed system usually occur at various times in the life of the system rather than all at the same time.

Problem Statement

The River Town City Council is faced with the problems of increasing its water supply during drought periods, increasing the availability of electric power that it now supplies from a city-owned steam plant, increasing the amount of recreational facilities available to its residents, and finding a way to decrease the amount of flooding that occurs in the section of the city along the river front. One alternative under consideration is a multiple purpose reservoir that can be used for recreation, hydropower generation, water supply, and flood control. It is estimated that the reservoir will cost $40.5 million to construct; will have annual operation, maintenance, and repair costs of $0.5 million; and annual benefits of $0.9 million for recreation, $0.7 million for power, $0.4 million for water supply, and $1.0 million for flood control. The reservoir is assumed to have a 50-year life and the benefits begin the first year. It is further assumed that a 5 percent discount rate is appropriate. The situation is summarized in Figure 9.1 in which the arrows above the horizontal line represent benefits and the arrows below the line represent costs. The benefits for the four purposes total $3.0 million per year. The usual convention is to assume that the construction cost occurs at the beginning of the first year and that the annual costs and benefits occur at the end of each year.

Since the reservoir has a long life, it becomes necessary to consider the annual operating costs and annual benefits when evaluating the system. Thus, there is a need to compare costs and benefits that occur at different times.

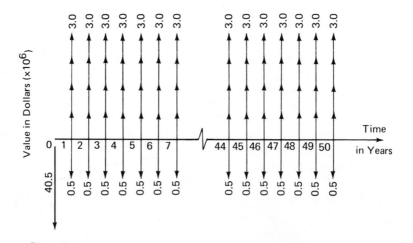

Figure 9.1

Time Profile of Benefits and Costs for Reservoir Example

Time Value of Money

The value of money is not fixed. It varies with time because of its ability to grow if profitably invested or to earn interest if placed in a savings account. A dollar today is equivalent to $1.05 one year from now if an investment opportunity is available that will earn 5 percent interest per year. If a person is given the choice of receiving $1 today or $1 one year from today, he would choose the $1 today because the $1 could be invested and earn interest. This concept of equivalence is important in economic studies that justify proposed systems because it allows values that occur at different points in time to be compared on a common time basis.

Narrowly defined, in a borrowing-lending situation, interest may be defined as money paid for the use of borrowed money. The rate of interest is the ratio, expressed as a percentage, of the interest payable at the end of a period of time, usually a year or less, and the money owed at the beginning of the year. Thus, if $5 is payable annually on a debt of $100, the interest is $5 and the interest rate is $5/$100 = .05, or 5 percent per year.

Although the interest rate may be loosely defined as any expression of the time value of money, a more precise definition distinguishes between interest rate and discount rate. Interest is the fee for borrowing money. A discount rate is the expression of the time value of money used in the equivalence computations. The discount rate should reflect the opportunity cost, which is equal to the return that would have been realized had the money been invested in other alternatives available to the individual. It may or may not be equal to the interest rate. If the individual can earn 8 percent interest on his money from some other source if the system is not constructed, then he should use 8 percent for the discount rate even though he can borrow money from a bank at 6 percent.

Interest is said to be compounded if interest is earned on interest. If interest is based only on the original amount for each period, the interest is said to be simple interest. Most economic studies are performed at compound interest wherein values are adjusted forward or backward in time using a compounding growth or discount factor.

Present Worth

One way of comparing costs and benefits is in terms of their present worth. This is done by computing the equivalent costs and benefits as if they were to all occur at the same time, such as at the beginning of the life of the system.

Using the equivalence concept for compound interest introduced above, note that the value today of $1 to be received one year from now is only $1/(1.05); i.e., this is the amount that would be worth $1 one year from now if invested at 5 percent interest. The value today of $1 to be received two years from now can be determined by computing its value one year from now, which is $1/(1.05); i.e., the value that must be invested at 5 percent interest one year from now to be worth $1 one year later, which is two years from now. The value of $1/(1.05) today is then $1/[(1.05)(1.05)] = $1/(1.05)^2. This can be generalized to

$$PWSP = \frac{F}{(1 + i)^n} \tag{9.1}$$

where $PWSP$ is the present worth of a single payment F that is n years from now when the discount rate is i.

The present worth of a series of equal annual payments, or receipts, such as the annual operation, maintenance, and repair costs of the reservoir, is the sum of the present worth of each cost. Therefore,

$$PWUS = \frac{A}{(1 + i)} + \frac{A}{(1 + i)^2} + \cdots + \frac{A}{(1 + i)^n} \tag{9.2}$$

where $PWUS$ is the present worth of a uniform series of A per time period, i is the discount rate, and n is the number of time periods. Equation 9.2 can be simplified to

$$PWUS = A \left[\frac{(1 + i)^n - 1}{i(1 + i)^n} \right] \tag{9.3}$$

The term in brackets in Equation 9.3 is known as the present worth uniform series factor and can be denoted as $(PWUS, i, n)$. Its value has been tabulated for various combinations of i and n in Appendix C.

The present worth uniform series factor is equal to 18.256 for $i = .05$ and $n = 50$. Therefore, the present worth of the operation, maintenance, and repair costs, $PWOMR$, of the reservoir is computed as

$$PWOMR = (\$0.5 \times 10^6)(18.256) = \$9.13 \times 10^6$$

Similarly, the present worth of the benefits, PWB, is computed as

$$PWB = (\$0.9 + \$0.7 + \$0.4 + \$1.0)(10^6)(18.256) = \$54.77 \times 10^6$$

Hence, the present worth of the total costs is $\$40.5 \times 10^6 + \$9.13 \times 10^6 = \$49.63 \times 10^6$ and the present worth of the benefits is $\$54.77 \times 10^6$. The net present worth is then equal to $\$54.77 \times 10^6 - \$49.63 \times 10^6$ and is equal to $\$5.14 \times 10^6$. This is sometimes referred to as the present value of the net benefits.

Annual Costs

Another way of viewing the value of a system is to consider its value on an annual basis. If Equation 9.3 is solved for A:

$$A = PWUS \left[\frac{i(1 + i)^n}{(1 + i)^n - 1} \right] \tag{9.4}$$

where the term in brackets in Equation 9.4 is known as the capital recovery factor and is used to convert a single value at the present time into an equivalent series of equal annual values. The capital recovery factor is equal to 0.05478 when $i = .05$ and $n = 50$. Therefore, the $40.5 million construction cost of the reservoir is equivalent to a series of annual costs C, computed as

$$C = (\$40.5 \times 10^6)(.05478) = \$2.18 \times 10^6$$

The total annual cost is then the sum of the annual operation, maintenance, and repair costs and the annual costs for construction, which is $0.5 \times 10^6 + \$2.18 \times 10^6 = \$2.68 \times 10^6$. Hence, the annual net benefits are determined by $3.0 \times 10^6 - \$2.68 \times 10^6$ and are equal to 0.32×10^6.

Price Changes

Price changes may be divided into two components. The value of money may decrease (inflation) or increase (deflation) with time and changes in the supply and demand for specific commodities with time may cause their value to change relative to overall price levels. An economic study must use commensurable value units. The most satisfactory value unit is money expressed in constant dollars. Dollars spent at one date may be transformed to constant dollars at another date by use of a price index.

A price index is the cost of a preselected group of items called a bundle of goods expressed as a percentage of the cost of the same items at some base date. *Engineering News Record (ENR) Construction Cost Index* is one of the most widely used indexes. This index is computed from the average cost of heavy construction consisting of fixed quantities of common labor, cement, steel, and lumber for 20 cities in the United States. The index was 100 for the base year of 1913. It equaled 797 in 1959 and 1,270 in 1969. Other price indexes are also available and current values for most cost indexes associated with construction projects are

, published four times a year in the Quarterly Cost Roundup of the *Engineering News Record.*

The use of an index assumes that cost varies with time in the same manner as the value of the items on which the index is based. Before a price index is used, the cost items under study should be compared with the cost items on which the index is based in order to determine whether their respective patterns of cost change are similar.

In estimating future costs for economic studies, prices under normal or average conditions should be used. This normalized price is obtained by averaging the price in constant dollars over a number of recent time periods to prevent project feasibility from depending on short-term market abnormalities.

The current normalized price in constant dollars needs adjusting only when the prices of goods or services are expected to change relative to the general price level. This adjustment involves multiplying the future cost by the ratio of the present to the future value of money. The future cost is determined by multiplying current cost by the ratio of the future to present index for the item. If the present value of the item is $100 and the value of money changes from 110 to 120 (as determined by a general price index) and the value of the item under consideration changes from 115 to 160 (as determined from a cost index for this item), the future cost in constant dollars is $(160/115)(110/120)(100)$, or $127, which is the value that should be used in the economic study. Uncertainty in predicting future changes in indexes usually limits this type of adjustment to less than 10 years into the future.

Construction cost indexes are particularly useful for estimating current construction costs of system components from historical data of completed systems. These estimates can then be used in the economic analysis.

9.4 ECONOMIC CRITERIA

Five economic criteria that can be used to rank alternatives will be explained here. The criterion chosen for selecting the alternative will depend on the objective of the owner.

Net Present Worth

The net present worth is an economic criterion that can be used to measure the effectiveness of alternatives. Calculating net present worth from a time profile of costs and benefits is straightforward. The following summarizes the procedure:

1. Discount each present worth to the same base year because sums of money at different times are different economic goods.

2. Use the same discount rate for all present values.

3. Base the present worth of each alternative on the same period of analysis. This may be done by evaluating the cost of extending the shorter-lived alternative to the entire period of analysis.

The net present worth of the River Town reservoir was shown to be 5.14×10^6.

If the choice of alternatives is simply how large to make the reservoir, then the maximum net present worth is obtained when the incremental present worth of the benefits is exactly equal to the incremental present worth of the costs. This occurs when the present worth of increasing the size of the reservoir one unit of size is just equal to the present worth of the costs associated with this unit increase in size.

Annual Equivalent Value

The basic idea here is to reduce the net present worth to a period (or annualized) basis. This, of course, shifts focus from the size of the system to the length of its expected life. The method for converting the net present worth value to an annual equivalent value is to multiply the net present worth value by a capital-recovery factor, as illustrated above for the River Town reservoir.

Now alternatives with different life spans can be evaluated and the equivalent annual values may serve as a criterion for ranking the alternatives. If the choice of alternatives is simply how large to make the reservoir, then either the present worth or equivalent annual values may be used as the criterion.

Benefit-Cost Ratio

The benefit-cost ratio is the ratio of the present value of the benefits to the present value of the costs. Annual values can be used without affecting the ratio. For the River Town reservoir, the benefit-cost ratio is

$$B/C = \frac{\$54.77 \times 10^6}{\$49.63 \times 10^6} = \frac{\$3.0 \times 10^6}{\$2.68 \times 10^6} = 1.12$$

When comparing alternatives, the same period of analysis must be used if the benefit-cost ratio is computed using present values. If the annual values are used, then the periods of analysis do not have to be the same for different alternatives because the annual values are based on the assumption that the shorter-lived alternatives would be extended to the entire period of analysis.

Internal Rate of Return

The internal rate of return is the rate of return used for discounting such that the present value of all benefits exactly equals the present value of all costs. This value cannot be determined directly, but must be solved for by trial and error; i.e., different interest rates are tried until one is found that causes the present value of benefits to exactly equal the present value of costs. In addition to being computationally cumbersome, there may be more than one solution for certain cases.

Least Cost

In many cases, it may not be possible to express the benefits in dollar terms. One way of ranking alternatives is to rank them by the cost of providing the same level of benefits.

9.5 CONSIDERING NONCOMMENSURATE VALUES

In the above analysis, it was assumed that the benefits from providing recreation facilities, hydroelectric power, water supply, and flood control are commensurable; i.e., they are measured in the same units, which in this case are dollars. However, there may be no way of objectively assigning dollar values to items such as recreation. In many cases, the evaluation must be made on subjective value judgments.

There are three ways in which subjective value judgments can be incorporated into the analysis. One procedure is to assign a value to the item based on the value judgment of the owner. This can be done by choosing a dollar value for each unit of the item, as was done above. A second procedure for incorporating value judgments is for the owner to specify some minimum quantity of one item. This minimum quantity then becomes a constraint that must be satisfied. It is not unusual for the owner to specify that a certain quantity of water supply must be provided by the reservoir and then the optimum reservoir size must be determined based on maximizing net present worth subject to this constraint. A third procedure is to develop a goals-achievement matrix (Hill, 1968) and subjectively choose the alternative.

Table 9.1 is a typical goals-achievement matrix that might be prepared for the River Town City Council. Where possible, the benefits and costs associated with each objective of each alternative have been expressed in terms of dollars. It is almost impossible to express the value of some items, such as natural woodlands or a wildlife refuge, in dollar terms. However, the impact of the alternative on these items is shown in the matrix in terms of units that are more easily expressed and understood, and the City Council must now directly consider this impact when selecting the alternative to be implemented. If there were some way to objectively assign a relative weight to the cost and benefit associated with each item, they could all be combined into one criterion function that could be maximized. Since this cannot be done at present, the decision-maker must subjectively make his decision based on his values for the contributions that each alternative makes toward fulfilling the stated objectives.

The goals-achievement matrix, as with most evaluation procedures, was designed to help compare and rank alternatives rather than test their absolute value.

9.6 FINANCIAL ANALYSIS

A system may be economically feasible but not financially feasible. In addition to the requirement that the benefits to the owner must be greater than his cost, there must be a plan for paying for the system. Few large systems are ever paid for from cash on hand. Most large systems are financed by loans that have to be repaid.

Bonds

Loans are often obtained through the sale of bonds. A bond is an instrument setting forth the conditions under which money is loaned. It consists of a pledge by a borrower to repay a specified principal at a stated time and to pay a stated interest rate on the principal in the meantime.

The interest rate that a borrower must pay on a particular bond issue is determined by competitive bidding. Interested buyers, usually

Table 9.1

TYPICAL GOALS-ACHIEVEMENT MATRIX FOR RIVER TOWN CITY COUNCIL

ALTERNATIVES		Water Supply	Hydroelectric Power	Flood Control	Recreation Facilities
A	Costs*	$0.4 × 10^6 and 500 acres of natural woodland will be removed from public access	$0.6 × 10^6 and will remove 1,000 acres of natural woodland from public access	$0.9 × 10^6 and will remove 2,000 acres of natural woodland from public access	$0.8 × 10^6 and 5,000 acres of a wildlife refuge area will be converted to recreational use
	Benefits*	$0.4 × 10^6 from sale of water and will support continued community growth for next 30 years	$0.7 × 10^6 in power revenue	$1 × 10^6 reduction in flood damages each year	Provides facilities to be used by 1.2 × 10^6 visitors per year
B	Costs*	$0.2 × 10^6	$0.9 × 10^6	$0.6 × 10^6 and two streets along river will have to be closed	$0.9 × 10^6 and remove 10 miles of river from use by those who like to canoe
	Benefits*	$0.5 × 10^6 from sale of water and will support continued community growth for next 40 years	$0.7 × 10^6 in power revenue	$1.1 × 10^6 reduction in flood damages each year	Provides facilities to be used by 2.1 × 10^6 visitors per year in water-based recreation

OBJECTIVES

* Dollar values are annual equivalent values. All values are the portion of total values for the alternative that is assigned to that objective.

large banks or related financial institutions, bid the amount they are willing to pay to secure the fixed sum of money on a fixed schedule. The borrower accepts the highest bid. The interest rate is calculated as the rate of return obtained by the investors on the face value of the bonds because the high bid hardly ever equals the face value of the issue.

Assume that the River Town City Council has decided to construct the reservoir discussed at the beginning of Section 9.3 and that the construction will be financed by the sale of bonds. The annual operation, maintenance, and repair costs can be paid out of yearly income from the reservoir. Therefore, bonds will have to be sold only to finance the construction of the reservoir. Since the construction cost of the reservoir is estimated to cost 40.5×10^6, the Council decides to sell 42×10^6 worth of bonds to make sure that they have enough money in case the cost estimate is low. They state that the 42×10^6 will be repaid at the end of 50 years and will pay 5 percent interest per year.

The highest bid they receive for the bonds is 41×10^6, which means they can have the use of 41×10^6 in exchange for a payment of $(.05)(42 \times 10^6)$, or 2.1×10^6, at the end of each of the next 50 years and a principal payment of 42×10^6 at the end of the 50-year period. The rate of interest that the city must pay on the amount of money borrowed (41×10^6) can be determined on the basis of the present worth of the two time profiles of money; i.e., the present value of 41×10^6 must be equal to the present value of 50 annual payments of 2.1×10^6 plus the present value of a 42×10^6 payment 50 years from now.

This condition may be written:

$$\$41 \times 10^6 = \$2.1 \times 10^6 (PWUS,i,50) + \$42 \times 10^6 (PWSP,i,50)$$

which must be solved by trial and error. The present worth of the amount received for the bonds is 41×10^6. To determine the present worth of the cost to the city, try $i = 0.05$. Then

$$\text{Present worth of cost} = \$2.1 \times 10^6 (18.256) + \$42 \times 10^6 (0.0872)$$

$$= \$42 \times 10^6$$

Try $i = 0.06$. Then

$$\text{Present worth of cost} = \$2.1 \times 10^6 (15.762) + \$42 \times 10^6 (0.0543)$$

$$= 35.38$$

The value of i falls between $i = 0.05$ and $i = 0.06$. Thus, by interpolation for i,

$$i = \left[0.05 + .01 \times \frac{41 - 42}{35.38 - 42} \right] = \left[0.05 + .01 \times \frac{-1}{-6.62} \right]$$

$$= 0.0515$$

Therefore, the city is actually paying 5.15 percent interest to borrow money.

Repaying the Loan

The city must provide a means of repaying the loan. The money for repayment may come partly from revenues received as a result of the sale of hydroelectric power and water and partly from tax revenues. Because these revenues are collected annually, the city needs to determine what the annual revenues must be in order to make the yearly payment of 2.1×10^6 for interest and still have enough left at the end of the 50 years to pay the principal of 42×10^6. In order to accumulate sufficient funds to redeem the bonds, annual payments must be made to a fund that, when invested at compound interest, will produce the required amount of principal when it becomes due. The time profile of the annual payments, A, as shown in Figure 9.2, must be equivalent to a final payment, F, at the end of the period.

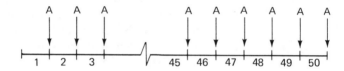

Figure 9.2

Time Profile of Payments to Sinking Fund

The value of the last annual payment at the end of year 50 is A. The value of the payment made at the end of year 49 has increased in value to $A(1 + i)$ at the end of the 50-year period. Similarly, the payment made at the end of year 48 has increased to $A(1 + i)^2$. Since the final sum, F, is equal to the future value of all annual payments,

$$F = A + A(1 + i) + A(1 + i)^2 + \ldots + A(1 + i)^{n-1} \quad (9.5)$$

where n is the number of time periods. This equation can be simplified to

$$F = A \left[\frac{(1 + i)^n - 1}{i} \right] \quad (9.6)$$

where the term in the brackets is called the uniform series compound amount factor. Equation 9.6 can be solved for A to yield:

$$A = F \left[\frac{i}{(1 + i)^n - 1} \right] \quad (9.7)$$

where the term in brackets is called the sinking fund payment factor. Using Equation 9.7, the annual payment to a sinking fund is

$$A = \$42 \times 10^6(0.00478) = \$0.2 \times 10^6$$

if the sinking fund money can be invested at 5 percent interest. Therefore, the income for the city must include at least 0.2×10^6 per year for the sinking fund plus 0.5×10^6 for OMR costs, plus 2.1×10^6 for the annual interest payment. If the annual income from the reservoir is not at least 2.8×10^6, then the city must provide money from some other source, such as taxes.

9.7 SUMMARY

All proposed engineered systems usually have to pass engineering, economic, financial, political, and social feasibility tests. These tests are applied to ensure that the system will be able to perform its intended function, that the benefits to be gained from the system are greater than the cost of implementing and operating the system, that the owner will be able to pay for the system, and that the individuals who will be affected by the system will respond favorably.

The purpose and scope of the planning are major factors in determining the planning horizon to be used in a study. There is considerable uncertainty associated with many of the benefits and costs of any system under study. The longer the period of analysis used, the more uncertain many of the costs and benefits become. Sometimes techniques, such as simulation (Chapter 8), are used to gain an understanding of the possible time profile of cash flows associated with the system.

The value of money varies with time because of its ability to grow if properly invested or to earn interest if placed in a savings account. Therefore, discounting factors must be used to convert costs and benefits to the same time period so that they can be compared. Along with the comparison of benefits and costs comes the desire to rank alternatives according to their desirability.

If all benefits and costs associated with the system can be expressed in terms of dollars, then criteria such as the net present worth, annual equivalent value, benefit-cost ratio, and internal rate of return can be used to rank the desirability of alternatives. The particular criterion to be used will depend on the objective of the owner. If the costs can be expressed in terms of dollars, but the benefits cannot, then it may be possible to rank alternatives according to the cost required to provide the same level of benefits. This criterion is usually satisfactory only for single purpose systems.

The economic ranking criteria were originally developed for single purpose systems that were independent of one another, and their use further assumes an unlimited capital available for investment. The ranking of interrelated systems under conditions of limited capital becomes a much more complex problem, as shown by Masse (1962).

When the benefits and costs cannot all be expressed in terms of dollars or any other common units, then the final selection must be made on the basis of the decision-maker's subjective judgment. This is usually the type of problem encountered in reality. Hence, we see that evaluation techniques are helpful in ranking alternatives, but the final selection depends on the values and objectives of the decision-maker, who is often the owner (individual, company, or government agency).

A financial analysis determines if money can be made available for system implementation after the owner has determined the alternative that he prefers.

If the proposed system passes the above feasibility tests, then the engineer must begin to plan the sequence and duration of activities necessary to construct the system. This process is discussed in Chapter 6. In addition, he must develop support organizations to perform the construction. These are discussed in Chapter 10.

9.8 PROBLEMS

P9.1. Equivalence calculations.

1. If a company puts $5,000 in a savings account that pays 6 percent compounded annually, how much will it have at the end of 10 years?

2. A company sets aside $5,000 a year of its profits for future expansion of the company. The money is put into a savings account that pays interest at 6 percent compounded annually. How much money will be available to the company for expansion 8 years after it starts this practice?

3. A concrete company will have to replace its mixing facilities at the end of the next 7 years. It is estimated that a new mixing facility will cost $40,000. How much must be put into a savings account each year in order to save enough money to pay cash for the facility if the money will earn 5 percent interest compounded annually?

4. If the company doesn't pay cash, it must borrow the $40,000 at 7 percent interest compounded annually. What would its

annual payment be in order to repay the loan in seven equal annual payments?

P9.2. A company has $1,200,000 to invest. One alternative is to invest the money in bonds that will pay 8 percent compounded annually. Another alternative is to invest the money in a power generating station. The generating equipment will last 50 years. The net income (income in excess of operating expenses) is estimated to be $10,000 the first year, $20,000 the second, $45,000 the third, $65,000 the fourth, $95,000 the fifth, $110,000 the sixth, $115,000 the seventh, $120,000 for the eighth, $125,000 for the ninth, and $130,000 for each remaining year of life of the station. Should the company invest in the power station? What would be the internal rate of return on this project?

✓ **P9.3.** A section of roadway pavement costs $5,000 a year to maintain. What immediate expenditure for a new replacement for the existing pavement is justified if no maintenance will be required for the first 5 years, $1,100 per year for the next 10 years, and $5,000 a year thereafter? Assume money to cost 6 percent.

P9.4. You are considering the improvement of some manufacturing facilities by purchasing one of three possible different machines, each with the same production capacity.

Machine A costs $30,000, has a life of 40 years, annual maintenance of $1,500, and salvage value of $5,000. Machine B costs $20,000, has a life of 20 years, annual maintenance of $2,000, and salvage value of $3,000. Machine C costs $10,000 and has a life of 10 years. Machine C is the cheapest and has the poorest quality control during manufacture. Therefore, annual maintenance costs may vary. Available information indicates that the probability of an annual maintenance cost of $4,000 is 0.5, of $3,000 is 0.3, and of $2,000 is 0.2. Machine C has no salvage value.

Use an annual discount rate of 6 percent. Assume that initial costs, annual maintenance, and discount rates are constant throughout any period of time you desire. Show a decision tree with cost of all alternatives and indicate the most economical choice.

P9.5. A firm is considering building a high-rise building that it will finance through the sale of bonds. It issues 30-year bonds with a face value of $4,000,000 bearing interest of 5 percent payable annually. The firm finds, however, that it can sell these bonds for only $3,900,000. What is the yearly rate of interest that the firm must actually pay for the funds received?

P9.6. A dam is to be constructed to provide a flood storage reservoir for a certain flood control project. Following are cost-benefit data for reservoirs of several sizes at the proposed dam site.

RESERVOIR VOLUME (ACRE-FEET)	INITIAL CONSTRUCTION COST ($ \times 10^6$)	AVERAGE ANNUAL OMR COST ($ \times 10^6$)	AVERAGE ANNUAL BENEFITS ($ \times 10^6$)
50,000	4.5	0.032	0.317
100,000	5.0	0.079	0.761
150,000	8.0	0.127	1.078
200,000	14.0	0.143	1.269
250,000	22.0	0.190	1.332

Determine the project size (i.e., reservoir size) that yields:

1. The maximum value of net benefits
2. The maximum benefit-to-cost (B/C) ratio

Assume that the project life is 50 years and that the appropriate discount rate is 6 percent.

P9.7. A Tollroad Authority has decided to build a new toll bridge across a river at a cost of $15 million. The bridge is to be paid for by selling bonds that are to be retired at the end of 40 years. The bonds are to pay 5 percent interest, payable annually. A sinking fund is to be established to pay interest on the bonds and to retire them at the end of 40 years. The sinking fund is to be invested at 6 percent interest, compounded annually, and is to be established in 40 equal installments, 1 year apart, the first installment 1 year after the bonds are issued. Assume that the bonds are sold at face value. How much should the toll per vehicle be if it is estimated that 50 million vehicles per year will use the bridge? The annual operating and maintenance costs for the bridge are $500,000.

P9.8. What factors would you consider in evaluating the feasibility of locating and constructing a new highway through your city? How would you develop a goals-achievement matrix to evaluate the effects of the new highway? (Hint: See Hill, "A Goals-Achievement Matrix for Evaluating Alternative Plans.")

10

PROJECT MANAGEMENT

This eight-story steel frame of the St. James Apartments in Treasure Island, Florida, took only 12 working days to erect. Shorter overall construction time with attendant lower labor costs and earlier occupancy are all economic pluses. (Courtesy of the Civil Engineering Magazine, American Society of Civil Engineers)

10.1 THE NATURE OF PROJECT MANAGEMENT

Developing and implementing a project requires that a variety of resources be identified, mobilized, and applied effectively to work tasks throughout the project life. Six basic project resource types can be readily identified; namely: labor, equipment, materials, cash, information, and decisions. The physical prerequisite for beginning a work task activity is that the necessary labor, equipment, and materials be available. Cash resources provide the feasibility environment for the physical resources to be provided. The information and decision resources provide the technological and conditionally permissive environment to be established in which the activity can actually be performed.

The application of these project resources over time can be considered as project flows. A variety of interrelated and dependent flows are generated because of the complexity of a project, so that project management can be viewed as the manipulation of networks of interacting resource flows. Initiating, speeding up, slowing down, and terminating an activity represent decisions relating to the rate at which a resource type flows in the project and represent management's involvement in the activity.

At the project activity level the basic resource flows involve the discrete movement of isolated entities. In bricklaying, for example, a certain number of trucks, loaders, hoist, masons, laborers, and brick pallets are involved. As these discrete flows are viewed from higher levels of the decision hierarchy within the project organization, they tend to integrate into continuous flows of men, materials, and resources of various types.

Project management is concerned with the dynamic commitment of resources to ensure completion of the project. It, therefore, must consider the determination, procurement, allocation, and utilization processes. The general project organization and the system of project administration and supervision must be devised with these primary objectives in mind. Project management must perform the basic management functions of organizing, staffing, deciding, directing, monitoring, and controlling all aspects of the project implementation. These functions are required for all projects whether implementing a design process, operating an agency, laying out an office, or constructing a building or engineered facility.

Project management is readily identified with construction. Construction projects can be broadly classified as building, engineering, or industrial depending on whether they are associated with housing, social works, or manufacturing processes. Engineering projects are often located in remote and rugged areas whereas building projects may be located in the midst of a large and populous city. Local geographic and environmental conditions play dominant roles in project definition and determining construction methods and activity sequences. In many cases, establishing lines of access, supply, and communication require major engineering efforts. It may even be necessary to develop and build site support facilities such as housing, quarries and aggregate crushing plants, concrete and bituminous mix plants, cable ways, and the like. City building projects, on the other hand, require a different emphasis, especially for problems raised by access and supply constraints imposed by city congestion and limited on-site storage facilities. All the necessary activities and auxiliary works indicated above need to be considered and their effects included in the project definition and CPM model.

In all cases, construction projects require reaching out from the office and design environment into field execution and management. Project management in the field requires establishing a field organization to direct and monitor field works and to ensure that the project progresses smoothly. It must be concerned with procuring, supplying, and maintaining all the resources required for the project.

The personnel organization must include field supervisors to direct the day-to-day activities at the site. In addition, office personnel will be required to record and evaluate field reports, handle financial payments to labor and subcontractors, and help monitor the support activities necessary for construction of the project.

Because there is a wide range of operations and procedures, work sites are temporary and often remote, the local site management rarely has full control of policy and finance, and can never be self-sufficient. In addition, since many of the site personnel are transitory, the engineer must plan how personnel will be organized and information will be processed so that the project can be supervised.

Also, monitoring and control systems to control the flow of material, equipment, labor, and finance must be established. Management is also concerned with the form of the information system and the nature of the decision-making process as determined by the field organizational structure and the owner-contractor relationship.

The information system must be planned and organized so that it allows the engineer to determine the state of the construction process at different points in time. He must plan what kinds of activities are needed and how often they are to be monitored and reported. He must also specify what kind of data is to be reported for each activity. This information, which defines the state of the construction process, becomes very important when a change in the environment necessitates revising the construction schedule and using more men, equipment, or materials. A flood may delay an activity; a strike may prevent the use of labor for a period of time; equipment may break down; or material may not be delivered on time. Any one of these events will force the engineer to make a series of decisions about how to reschedule activities, how fast to perform the activities, whether to use more men and equipment, and whether there is a way of completing the project in the specified time.

10.2 THE DECISION CYCLE

Project management is primarily concerned with decisions affecting the rate of flow of resources to the project, a fact that is more readily seen in construction projects than in others. Construction is, by its very nature, a complex undertaking requiring the organization of a wide spectrum of diverse elements. One of the major factors that sets the management of construction projects apart from project management in other fields of endeavor is the high variation in the efficiency of performing the construction operations that form the components of the project. This is a function of the highly uncertain character of the construction environment. Environmental considerations (temperature, precipitation, and site conditions) interact with economic considerations in providing the dynamic setting within which a construction manager must make his decisions. These environmental factors obviously influence worker efficiency and thereby control the progress of the project. Experience in handling decisions under such circumstances is one of the successful project manager's basic managerial tools.

The elements that structure the manager's reaction within the construction environment appear to be basic to the management process (Halpin and Woodhead, 1971). The elements are:

1. Postulating a problem and planning for its accomplishment,

2. Specifying the monitoring variables that reflect on the state of system as it responds to the random effects of environmental disturbances,

3. Determining the system variables under the control of the manager by means of which the manager provides corrective action.

The decision process reduces to the basic management functions as shown in Figure 10.1.

This process illustrates the iterative decision loop that appears to be basic to all construction management situations. This loop consists

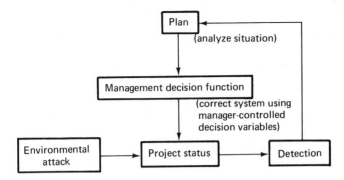

Figure 10.1

Management Functions in the Decision Process

of the following functions:

1. *The Planning Function*—defines a conceptual model of the task to be undertaken or the problem to be solved; it provides the logic that defines the problem. In construction, this often takes the form of defining the network model of a project.

2. *The Management Decision Function*—this function represents the means whereby the manager can manipulate and select among the alternatives open to him and exert his will on the solution of the problem. This decision function must be expressed in terms of the variables available to the decision-maker and must then be transformed into terms appropriate to the definition of the plan. In construction, the choices relating to labor management, for example, include the increase of crew size, decision to work extra days, and so on, and can be converted into time and cost parameters for the project and incorporated into the project network model. The goal of the decision function is to initiate a course of action to achieve a desired end.

3. *The Environmental Attack*—no course of action in construction is shielded from the effects of an environment that often strongly influences the desired outcome of the decision process. Environmental variables are beyond the control of the manager and affect the planned rate of progress. The environment is characterized by variables such as rainfall, temperature, and economic index.

4. *Detection Function*—the interaction of the decision function and the effects of the environmental attack both acting on the original plan produce a *project status* or state that must be de-

tected and monitored by some relevant management information system. In construction, this becomes a reporting function focusing on current and cumulative cost and time attributes of the system.

This management decision cycle can be readily identified and applied to many construction management problems. These problems are all characterized by the necessity to react to difficulties produced by an unpredictable environment on a planned course of action.

Historically, the construction industry has adopted a hierarchical structure as an expedient in structuring its activities. From the point of view of the construction company, this hierarchy begins at the level of the foreman or superintendent and ascends to the owner or president of the company. Many management viewpoints are possible even when basically the same situation is involved. For instance, the foreman may be interested in efficient use of crew and equipment for a given process, while the project management is interested in leveling resources across the job and organizing supporting activities such as procurement and payroll. The company president, on the other hand, may be interested in labor, equipment, and capital investment ratios. Whatever the management viewpoint considered, the basic management functions illustrated in Figure 10.1 and discussed above must appear in any decision-making management situation.

10.3 PROJECT RESOURCE FLOWS

The basic component in project resource flow is the requirement of each project activity. The rate at which resources are applied to each activity over its duration is a management function that directly influences project resource flows. Although constraints on the rate at which project resources are applied may exist at the project level, resource usage is determined at the activity level.

Determining the magnitude and duration of activity resources requires the following steps in the management process:

1. Each project activity requires a certain amount of work to be done, which must be evaluated or *estimated* in terms of a physical unit of measure. These units may be, for example, square feet of masonry wall, cubic yards of concrete, etc. These estimates are often surprisingly accurate but may be incorrect due to human error or to actual variations in the field. The activity is completed when the actual total pacing quantity of work is completed. In this way the activity work quantity

can be linked to and thought of as an account of work units that must be reduced to complete the work to be performed.

2. In order to perform the work on the activity, management must allocate the necessary resources. This normally takes the form of providing a labor work force with supporting equipment. Thus, for example, erecting a brick wall requires bricklayers, laborers to erect scaffolding, to position brick pallets, and to maintain mortar supplies. In addition, a hoist operator may be needed as well as truck drivers to deliver the necessary brick pallets, sand, and cement. The identification of the various trades, the number and mix of workers involved defines the activity crew profile. The definition of the crew profile and equipment establishes the rate at which management intends to approach the activity.

3. The actual effective rate at which the activity is performed depends on the productivity of the applied resources, since only the output of these resources completes the activity. The crew profile must be so balanced that the critical trade or equipment that defines progress is fully supported. In bricklaying, for example, any delay in supplying either bricks or mortar reduces productivity and hence prolongs the activity duration and the period during which the flows of labor, material, and equipment must be maintained.

4. The duration of activities change because of incorrect estimates, low productivity, and environmental attack. Consequently, if activity durations are critical and must be maintained, management must continually modify the rate at which resources are applied. This may require providing additional bricklayers and larger on-site inventories of bricks, cement, sand, and so on.

5. Finally, work on each activity is conditional on the completion and status of other project activities according to the project logic and network model.

The variations introduced in each of the above steps mean that time-varying levels are associated with the activity unit accounts. In addition, time variance occurs because completion of work upstream in a sequential network of work activities generates a capacity for work in certain downstream or logically subsequent work accounts. For instance, an amount of excavation permits subsequent work on a given amount of form work and concreting. This effect is analogous to that found in a water resource system in which release of water at upstream reservoirs determines river stage and reservoir levels at downstream locations. In the water basin, the problem is to route water through the system in an optimal way so as to maximize its productive effect and minimize backup and flooding that cause damage. Similarly, in the construction situation, the problem is to optimally route the resource flows mentioned above through the project system so as to maximize project productivity while minimizing costly delays and idleness of the routed resources (Halpin and Woodhead, 1972).

Rates are essential to the concept of flows since they reflect the dynamic effect of reducing account levels to zero based on some function. The discrete flow models discussed above model a situation at the micro level of hierarchy and their integration results in continuous system flows. Because continuous flows are made up of discrete flows, discrete flow system outputs can be aggregated to provide input to the productivity functions that reduce work activity account levels to zero.

These continuous rate values represent management's policy on the working of various activities. These rates are limited by

1. Available labor and other resources,

2. Efficient crew sizes,

3. Limitations of site accessibility,

4. The amount of potentially available work required to efficiently initiate a given activity,

5. The progressive availability to the activity of work to be accomplished.

For example, the rate of bricklaying for a building project depends on the number of floors currently available and scheduled for masonry work. Several approaches of varying complexity and sophistication can be used in applying the diminishing concept of rates to flows (Forrester, 1961).

In order to consider the nature of flow rates and their interaction with levels, consider the diagram in Figure 10.2. This diagram represents the account concept in terms of a tank of fluid with a given Level A.

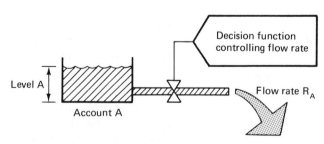

Figure 10.2

Diagram of Rate and Level Concepts

The valve located in the outlet line represents the application of some decision function that controls the outflow rate, R_A.

The simplest decision rule to use in controlling the flow rate would be to set the valve at one position resulting in a constant rate, R_A = constant. This is the same as diminishing the work accounts described above, using a constant or average crew rate until the account is reduced to zero. This results in a linear diminution that normally is assumed to be the average of several varying rates that accrue from skill level and environmental effects.

Another rate concept that is more complex and not normally considered when making time estimates for the duration of an activity or flow is that of a time varying rate. This time varying rate can be applied in a number of ways. For instance, R_A can be a function of a level that itself is time varying. That is, R_A = (level A)/P_A where P_A is a policy constant or function that reflects the priority value for completing the work. In this case, as the level of A varies downward over time, the rate will also vary. The rate might be a function of both level A (the source of the flow) as well as level B where B is the destination of the flow. This may be expressed as

$$R_A = \frac{\text{level } B - \text{level } A}{P_B}$$

where P_B again is a policy constant or function. For example, this situation is relevant when storage constraints dictate a reduction in the fabrication rate of precast members because of a slowdown in placement rates.

For a particular project activity the various resource flows can be modeled as shown in Figure 10.3. Initiating work on an activity is conditional on technological prerequisite conditions being satisfied as well as on the availability of a certain minimum level of materials, equipment, labor, and cash. This minimum level is called the "planning" level and reflects the fact that a certain quantity of resources must be available before it is feasible to begin work. Figure 10.3 illustrates the dynamic functioning and interaction of the project resource network flows necessary for each project activity and suggests management functions and the need for project organization.

10.4 PROJECT ORGANIZATION AND INFORMATION FLOW

In many projects, large numbers of workers are involved with a variety of work tasks over prolonged periods of time. The management and control of these projects require the definition and staffing of a project organization. Although the essential focus of management is on

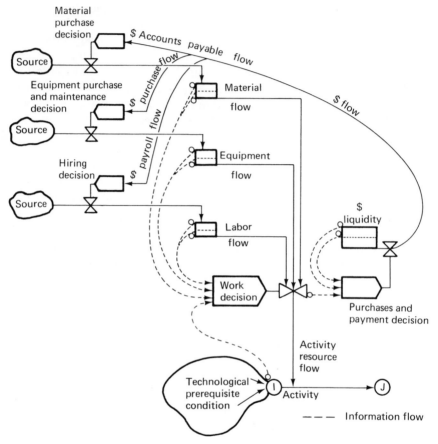

Figure 10.3

Project Activity Resource Flows

the actual technical and managerial direction of the project, a considerable organizational structure is required to acquire and process information relative to the project to insure that some dominant and general objective, such as profit, is achieved.

Project management requires that information be acquired, accumulated, and used for both normal handling of the project and as the basis for decisions, payments, and establishing historical records. The information requirements can be related to the flows in the basic project resources and to determining the current project status and environmental conditions.

Many different project organizational structures exist depending on the contractural relationships that exist between the owner agency, professional consultants, and the contractor and subcontractors involved in the actual construction. In addition, the relative location of the head office and the project field site and the management experience of the project staff will profoundly influence the project organization.

Project organizational structures are built around the basic functional and specialized areas of engineering, estimating, procurement, coordination, cost accounting, and payroll. Their relationships within the organizational structure and to the information flow relating to the basic project flows can be readily identified. Figure 10.4 indicates the functional breakdown for the head office organization of a typical building contractor, and Figure 10.5, the matching field organization.

The informational requirements for labor include monitoring and documenting labor time actually worked on the site, preparing payroll, and assessing tax and insurance deductions. The construction and management requirements require determining the work force size as a function of time for each phase of the project. The informational flow commences with the trade foremen and timekeepers and is transmitted by the job superintendent to the payroll section of the head office.

The material procurement requirements are initiated in the estimating, coordinating, and scheduling sections of the head office. The actual material procurement process of seeking suppliers, quotations, delivery promises, and the like, is a vital life function of the project organization and is often staffed by senior executives of the organization. In the field, the material flow organization comprises delivery and stores clerks and allocation of materials by the job superintendent and trade foreman.

The payments for materials, labor, and equipment require a payroll and accounting section and, consequently, a focus on cash flow and the preparation of financial statements. The various documents and their preparation and use constitute the information flow.

Finally, the information relating to the current status of the project—activities due to start, just completed—plus their resource requirements, constitutes the information transfer that is vital for project management. This information enables decisions and policy to be formulated and executed.

In all cases, the information flows within a project organization require identifying the data required, designing documents, and staffing and training agents within the organization. Attention to detail, to the manner, quantity, and type of information transferred by project staff, and the manner in which it is transferred, and the format and the layout of the various documents to be used, constitute the design and nature of the information system.

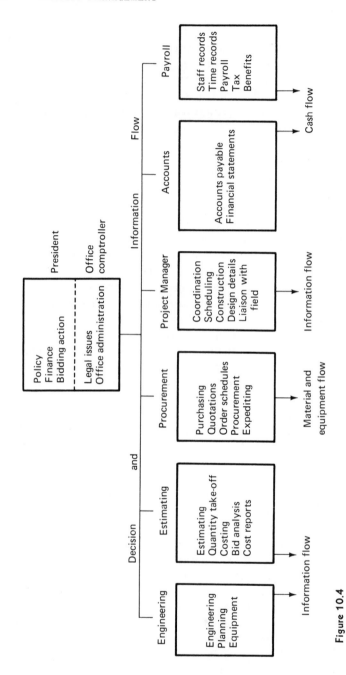

Figure 10.4

Head Office Organization (Functional Breakdown)

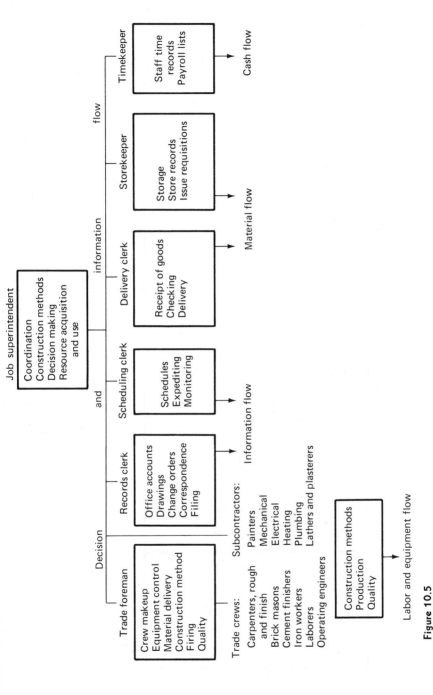

Figure 10.5

Field Organization (Functional Breakdown)

Procurement activities and durations (weeks) CPM	Design	Prepare Specifications	Estimate Quantities	Locate Suppliers	Award Bids	Order	Verify Quality and Details	Delivery Delay	To CPM Activity
	A	B	C	D	E	F	G	H	⊠
Item Agents	Engineers	Engineers	Estimators	Buyers	Legal	Accountants	Engineers	Expediters Delivery clerks	Project Engineer
1. Cement	0	1	1	1	4	2	1	½	B
2. Pumps	9	5	1	4	8	2	4	20	H
. . .									

Figure 10.6

Materials Procurement Chart

10.5 PROJECT RESOURCE FLOW MODELS

Simple project resource flow models can be derived from the CPM model if each activity is labeled with manpower and equipment requirements as well as its duration. Time cut-sets then enable resource profiles to be developed, as shown in Figure 6.8. In this way, both planned and actual resource flows can be portrayed.

In order to model the material procurement process, additional activities can be added explicitly to the CPM model or handled through the use of a supplementary Materials Procurement Chart and continuously updated *milestone nodes*, as shown in Figures 10.6 and 10.7.

Any alterations in the availability of labor and equipment, or delays in procuring materials, directly affect resource flows and will require a re-analysis of the project CPM model to determine their impact. It may be necessary to reschedule the project using a limited cut-set value of a resource following the concepts developed in Chapter 6. If this procedure is undesirable, management may need to implement a forewarning monitoring system so that procurement delays can be rectified or minimized. Finally, it may be possible to accept the current delays because at a later stage management may be able to regain the planned schedule using compression calculations with implied increased rates of resource allocations, as shown in Chapter 6.

Once the schedule of activities has been decided by locating each activity in calendar time, it is possible to determine the cash flow for the project. In many cases, the construction work is carried out under contract, with monthly payments based on the work that has been done in the field. The contractor performs work in the order specified by the CPM model under the supervision of the project engineer. Considering cash as a resource permits an analysis similar to that used previously for manpower resources. In addition, since the contractor must invest his own money to initiate construction before he is entitled to payment, he is vitally concerned with his own cash flows and overdraft requirements.

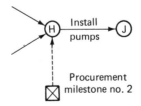

Figure 10.7

Procurement Milestone

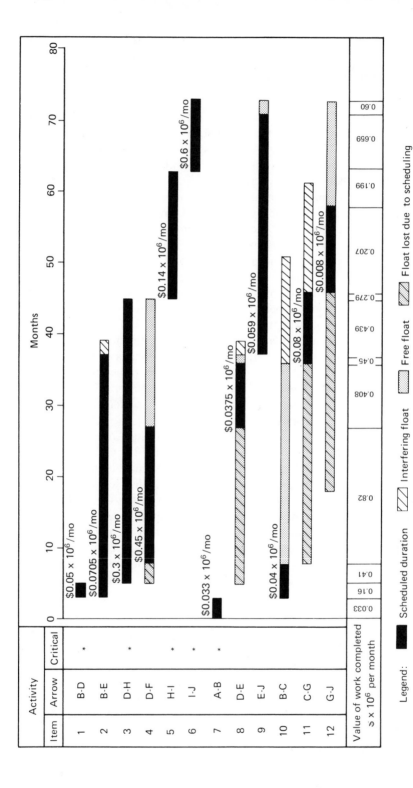

Figure 10.8

Bar Chart for Cash Requirements for Labor Constrained Schedule for Reservoir Construction

Legend:
■ Scheduled duration
▨ Interfering float
▥ Free float
▧ Float lost due to scheduling

Activity		Critical
Item	Arrow	
1	B-D	*
2	B-E	
3	D-H	*
4	D-F	
5	H-I	*
6	I-J	*
7	A-B	*
8	D-E	
9	E-J	
10	B-C	
11	C-G	
12	G-J	

Value of work completed $ x 10^6 per month

For example, Figure 10.8 shows a bar graph form of scheduling the reservoir project of Section 6.4 within labor constraints as shown in Figure 6.9. Table 10.1 summarizes pertinent data that includes the cost of

Table 10.1

SUMMARY OF ACTIVITY DURATIONS, COSTS, AND
BEGINNING OF MONTH STARTING TIMES

ACTIVITY	DURATION (MONTHS)	COST (DOLLARS)	STARTING TIME (MONTH)
1	2	0.1×10^6	4
2	34	2.4×10^6	4
3	40	12.0×10^6	6
4	20	9.0×10^6	9
5	18	2.5×10^6	46
6	10	6.0×10^6	64
7	3	0.1×10^6	1
8	8	0.3×10^6	29
9	34	2.0×10^6	38
10	5	0.2×10^6	4
11	10	0.8×10^6	37
12	12	0.1×10^6	47

accomplishing the various activities in the project. As work is conducted on each activity, a cost is incurred to the contractor. However, the contractor is paid only on completion of certain phases of the project or on some schedule that is based on completed work. Thus, the contractor must anticipate the income versus the costs incurred on the project in order to determine the financing that may be required.

For the particular project shown in Figure 10.8, assume that the costs are incurred uniformly over the period for each activity. These are shown above the bar in units of dollars per month. Of course, if the level of effort is variable throughout the activity, it may be necessary to consider variations in monthly costs.

At any time during the project, a vertical line may be drawn indicating the end of a particular work period. The summation of costs incurred along this cut-set line represents the expenditures for that period. Thus, costs can be obtained for each work period, which in this case, is in months. Also, the schedule of income can be anticipated.

Figure 10.9 shows the schedule of accumulated cost plotted graphically over the duration of the project. Using this curve, the engineer can anticipate the project cash flow and his finance requirements.

The contractor, however, must provide his own financing both

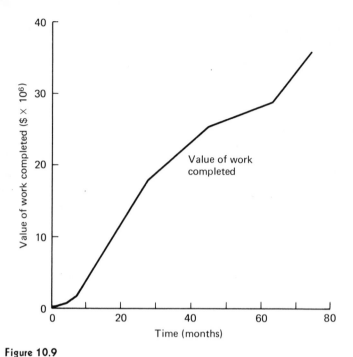

Figure 10.9

Owner's Project Cash Flow for Labor Constrained Schedule for Reservoir Construction

to initiate and to continue the project. As work progresses (usually monthly) he submits claims for payment that, if correct, are paid within the next month less a retainage (usually 10 percent). The retainage concept has been developed to insure that the contractor has an inducement to complete the project, at which time all the accumulated retainages are paid to him. Assuming the contractor is earning a net 5 percent profit, Figure 10.10 shows the calculation and graph of the financing requirements of the contractor for the first three months of the project. Notwithstanding payments to the contractor, his overdraft requirements are still increasing and may not stabilize for some time.

In the event that the contractor must reduce his overdraft requirements, or the owner is tardy in making payments, he must slow down his rate of working. The contractor may thus choose to shift activities within their total float as a means of reducing his cash requirements or he may reduce the rate of progress in all current activities and accept an extended project duration.

The preceding examples illustrate how the essential resource

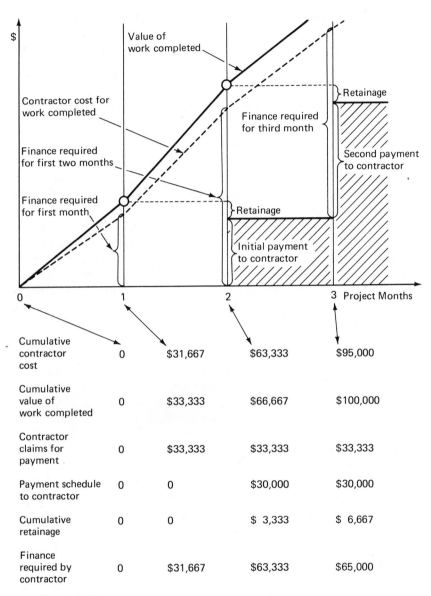

Figure 10.10

Financing Requirements of the Contractor for Reservoir Construction

flows of labor, equipment, material, and cash can be both modeled and managed. The project flows of information and decisions relate to the professional aspects of management and its organizational structure.

10.6 SYSTEM ANALYSIS METHODS FOR PROJECT MANAGEMENT

The concepts and methods of systems modeling and analysis can readily be applied to project management. Project definition and analysis in practice use linear graph models and critical path calculations, as indicated in Chapters 2 and 6. Cut-set concepts are in everyday use for labor, equipment, and cash flow analyses. Many computer programs are available for CPM calculations and are used by management in project control. Some have a problem oriented language whereby the normal professional statements required to define a problem or course of action are sufficient to activate the necessary data and algorithmic processes to solve the problem. In addition, a CPM based accounting and payroll system is available at the University of Illinois at Urbana-Champaign that integrates project information at the activity level, with the cash flow aspects of materials and equipment costing and labor payroll.

The decision focus of project management constantly calls for analyses based on alternatives and uncertainties and the decision tree modeling concepts of Chapter 7. Many cases also exist to which the linear programming models of Chapter 6 can be usefully applied. Situations where product mix questions must be resolved, or where alternative gravel pit locations are available and materials transportation is significant, are possible areas for systems analysis. Because, in general, the dominant effort in construction projects is directed toward the movement of materials, considerable economies may be possible by carrying out critical analyses on the transportation systems and activities performed in the procurement area and even on the site itself. The secondary and tertiary handling of materials requires labor, equipment, and time resources, and may be nonoptimal.

The simulation models of Chapter 8 may be used to investigate the productivity of repetitive or cyclic operations such as earth movement in highway construction or concrete pouring in dam construction. The influence of hoist capacity and the cost impact of material inventories can be modeled by simulation for construction activities, such as those required for extensive bricklaying projects.

As indicated in Figure 10.1, the essential ingredients in a management decision problem require the definition of a plan or basic logic plus the iterative determination of management action through decision variables. Many problems exist in project management and provided a systematic problem definition procedure is carried out as indicated in Chapter 1, a large variety of management decision variables and potential logics will be uncovered. These only await the engineer's perception and development of useful and practical management tools using systems concepts and methodologies.

10.7 PROBLEMS

P10.1. The contractor who is constructing the reservoir described in Section 6.4 had planned his work schedule as shown in Figure 6.8. However, at the end of 5 months he finds that he can hire a total of only 100 men because the rest of the available local workers have been hired to work on other projects.

1. What is the new minimum completion time and work schedule if an activity must be completed once it is started?
2. What is the new minimum completion time and work schedule if activities 2, 3, 8, 9, 10, and 11 can be worked on whenever there are men available from other activities?
3. What effect does the detail used in CPM modeling have on the options available to the construction foreman?
4. Discuss what other alternatives might be available to the contractor.

P10.2. The contractor involved in the reservoir construction problem has planned on receiving his payments for completed work as shown in Figure 10.10. However, as a result of a dispute, his second payment from the owner is delayed 1 month.

1. What is the financial impact of this delay if he continues work as planned?
2. What options are available to the contractor during the fourth project month for lessening this financial impact?

P10.3. A gravel company is under contract to supply aggregate for a dam construction in a remote location and some considerable distance along a poor quality access road. It operates and maintains its own large fleet of delivery trucks and is concerned about the deterioration and the frequency and amount of maintenance costs for operating its truck delivery fleet.

The company management is considering the possibility of financing a road improvement plan for the access road as a means of eliminating certain types of deterioration and reducing the frequency of maintenance work on its truck fleet. How would you go about establishing the maximum road improvement investment the company could tolerate?

P10.4. Establish the organization and operation policies used by a company you are familiar with for handling orders, inventory, and factory deliveries to its warehouse. Model the document and information flow systems and identify management decision problem areas. Can you develop any relevant decision models for the problems?

P10.5. The daily labor allocation problem is faced by project managers whenever planned labor resources are not met on the construction site. Some activities may be worked with reduced crew profiles between the trades involved if minimum requirements are met; others may be postponed. In some cases, available labor can be used to accelerate progress on "favored" activities.

Develop an integer linear programming model for the optimal allocation of available labor (from n trades) to work the scheduled m activities. What would be suitable objective functions for your model?

Use the following notation:

a_{ij} = number of trade j workers planned and scheduled to work activity i to make 1 day's progress in 1 day;

x_{ij} = number of trade j workers available and allocated to work activity i;

c_j = number of trade j workers actually reported for work this project day;

d_{ij} = decision variable to work activity i with trade j; $d_{ij} = (0,1)$ if activity i (is not, is) worked by trade j;

l_{ij}, h_{ij} = management defined range limits (low, high) for the number of trade j workers for activity i if it is worked.

In addition, to ensure reasonably compatible crew profiles among the trades working an activity, management defines a range of ratios (r_{low}, r_{high}) between the various trades (see Figure P10.5). Usually this can be done by referring to a specific trade in each activity that basically determines the productivity output for that activity.

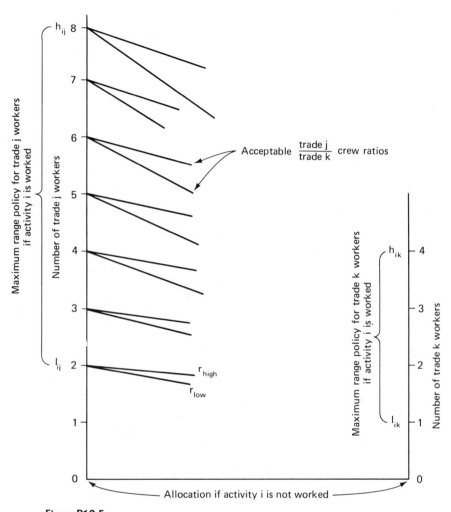

Figure P10.5

Acceptable Trade j,k Ranges and Rates for Activity i

P10.6. Determine the contractor's overdraft requirements over time if the reservoir project of Sections 6.4 and 10.5 is constructed according to

1. The earliest start times.
2. The latest start times.

3. What is the present worth cost to the contractor of financing the project for (1) and (2) if he must pay 6 percent interest on his overdraft? Assume that interest for 1 month is $\frac{1}{12}$ of what it is for 1 year and that the interest payments are made at the end of the year.

4. Discuss the significance of the results in (3).

P10.7. The toll bridge in Problem P9.7 has been in operation for 5 years. The actual traffic volume that uses the bridge is only 85 percent of the projected volume; hence the Toll Authority has not been able to make its total payment each year into the sinking fund. The Toll Authority must increase its revenue or extend the length of the bond issue. For each 10 percent increase in toll charge, the volume of traffic will decrease by 1 percent. If the bonds have to be reissued, the interest rate will be 6 percent instead of the current 5 percent rate.

1. What toll would have to be charged in order to make up the back payments and pay off the bonds on time?

2. What would be the value and length of a new bond issue that would replace the old bonds without a change in toll charges?

3. Discuss the relative merits of the two approaches and a combination of the two.

P10.8. Visit a construction site and attempt to establish the nature of the field organization and management methods employed.

1. What is the purpose of a project management information system?

2. Identify the basic ingredients required for any project management information system.

3. Indicate how you would establish data collection points and an information flow organization so that management is supplied with the necessary information for project control.

4. Suggest a feasible model that management could use to control one particular aspect of the project.

11

STATE CONCEPTS
OF SYSTEMS THEORY

The Oroville Dam and Edward Hyatt Powerplant, located on the Feather River in California, was selected as the Outstanding Civil Engineering Achievement of 1969. The dam forms a key storage unit for the California State Water Project. Oroville Dam is one of the world's highest earthfill dams. Water conservation and storage is the primary purpose of the dam. The generation of hydroelectric power, flood control, and recreation are important secondary functions. In 1955, before the dam was constructed a record flood on the Feather River caused the loss of 38 lives and $100 million worth of flood damages. In 1964; an even larger flood was successfully controlled although the dam was only in its second year of construction and only partially completed. (Courtesy of the American Society of Civil Engineers)

11.1 INTRODUCTION

A full-life description of any real engineering system is beyond human capabilities because of the infinite complexity of physical materials and forces, and the nature of the environment in which the system is embedded. Nevertheless, engineers are called on to plan, design, build, monitor, and maintain systems that perform specific engineering and societal functions over long periods of time. They must, therefore, in particular cases, gain an understanding of a system, develop adequate descriptions of the system, its environment, and its use, and thus be able to identify the changing status and behavior of the system over time.

The description of a system is therefore intimately linked with the construction and characteristics of a system model. Selecting a specific set of variables out of the large number that may exist enables a specific system description to be developed. An entirely different selection of variables may permit another and entirely different system description. Clearly it is important to consider carefully the purpose of the system description and to include all the dominant variables that affect the problem under consideration. In addition, understanding or predicting the actual system behavior involves interpreting, measuring, and modeling the environmental inputs that tend to drive the system toward a new posture and response.

It is natural, therefore, to focus attention on determining those system variables that can uniquely and adequately describe the system status (i.e., state) at any time. The system behavior is then portrayed by the succession of states that the system assumes at particular points of time.

11.2 STATE CONCEPTS

The system variables that are used to define or describe the system state (status) at any time are called state variables. A particular system state is then identified by a set of instantaneous values of the state variables. If the value of one or more of the state variables is changed so that a new configuration can be recognized, then a new state exists. A transformation function defines the new state in terms of the old state and the changes in the values of the state variables. State concepts are demonstrated in the following examples.

Example 11.1   Mr. Keep R. Spendit has a checking account with the First World Bank and a charge account with the Kredit Kard Company. At the beginning of each month he receives a statement from the bank that indicates how much he has in his checking account. He also receives a statement indicating how much he owes the Kredit Kard Company. These two values, then, represent his financial state at the beginning of the month. If he does anything during the month to change one of these values, such as making a deposit or withdrawal at the bank or making a charge or payment with the credit company, his financial state will be different at the beginning of the next month. Therefore, Mr. Spendit's financial state at the beginning of the ith month, $\mathbf{S}_i$, is

$$\mathbf{S}_i = \begin{bmatrix} s_1 \\ s_2 \end{bmatrix}$$

where s_1 represents the amount in his checking account and s_2 represents the amount he owes. His financial state at the beginning of month $i + 1$, $\mathbf{S}_{i+1}$ is then

$$\mathbf{S}_{i+1} = \mathbf{S}_i + \mathbf{I}_i \tag{11.1}$$

where $\mathbf{I}_i$ is the input during month i and is defined by

$$\mathbf{I}_i = \begin{bmatrix} a_1 \\ a_2 \end{bmatrix}$$

where a_1 represents the change in the amount in the checking account and a_2 represents the change in the amount owed. Thus, his financial state depends on his previous state and the input supplied, as indicated by Equation 11.1 and Figure 11.1.

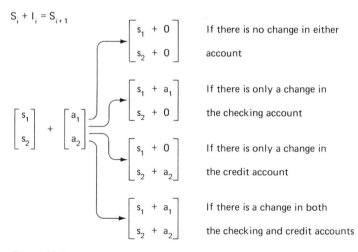

Figure 11.1

Schematic Representation of Transformation from State $\mathbf{S}_i$ at Time T_i to State $\mathbf{S}_{i+1}$ at Time T_{i+1}

In general, Equation 11.1 can be written as

$$\mathbf{S}_{i+1} = F_i\{\mathbf{S}_i, \mathbf{I}_i\} \tag{11.2}$$

where $F_i\{\mathbf{S}_i, \mathbf{I}_i\}$ is the transformation function that determines the succeeding state when the previous state and the input are known.

Example 11.2   The compound bar structural problems treated in Chapter 3 can readily be cast into the state concept form. Although the concept of changing state with changing input is important, in linear algebra cases (such as applies to linear elastic materials) it is usual to consider each new loading as an entirely separate case and thus to ignore incremental loading and incremental system changes. However, once the structure exhibits any nonlinear properties, the analysis must follow the state concepts as outlined in this section.

The following matrix algebra equations apply for the compound bar shown in Figure 11.2.

$$\mathbf{u} = \mathbf{Au'}$$
$$\mathbf{P} = \mathbf{Ku} = \mathbf{KAu'}$$
$$\mathbf{P'} = \mathbf{A}^T\mathbf{P} = (\mathbf{A}^T\mathbf{KA})\mathbf{u'} \tag{11.3}$$
$$\mathbf{u'} = (\mathbf{A}^T\mathbf{KA})^{-1}\mathbf{P'}$$

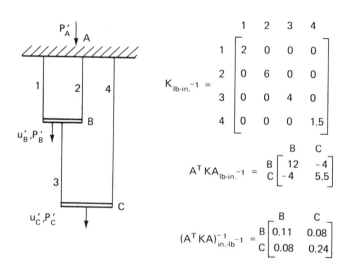

(a)   Schematic Diagram   (b)   Compound Bar Matrices

Figure 11.2

Compound Bar Example

The main solution effort is to determine the system response as given by the model displacement vector $\mathbf{u}'$ when the system is stimulated by the input joint loading $\mathbf{P}'$. Furthermore, all other system vectors; namely, $\mathbf{u}$ and $\mathbf{P}$, can be readily found as functions of $\mathbf{u}'$. It is natural therefore to consider the state vector as the nodal displacement vector $\mathbf{u}'$ so that

$$\mathbf{S}_i = \mathbf{u}'_i \tag{11.4}$$

Notice that the system state is directly determined by the applied loading; thus

$$\mathbf{S}_1 = \mathbf{u}'_1 = (\mathbf{A}^T\mathbf{K}\mathbf{A})^{-1}\mathbf{P}'_1$$

and

$$\mathbf{S}_2 = \mathbf{u}'_2 = (\mathbf{A}^T\mathbf{K}\mathbf{A})^{-1}\mathbf{P}'_2 \tag{11.5}$$

In order to formulate the system in a formal state system model, the second state, $\mathbf{S}_2$, must be obtained in terms of the first state, $\mathbf{S}_1$, and a system input. Therefore, if

$$\mathbf{P}'_2 = \mathbf{P}'_1 + \Delta\mathbf{P}'$$

then

$$\mathbf{S}_2 = \mathbf{u}'_2 = (\mathbf{A}^T\mathbf{K}\mathbf{A})^{-1}(\mathbf{P}' + \Delta\mathbf{P}')$$

$$\mathbf{S}_2 = \mathbf{u}'_2 = (\mathbf{A}^T\mathbf{K}\mathbf{A})^{-1}\mathbf{P}' + (\mathbf{A}^T\mathbf{K}\mathbf{A})^{-1}\,\Delta\mathbf{P}$$

or

$$\mathbf{S}_2 = \mathbf{u}'_2 = \mathbf{S}_1 + (\mathbf{A}^T\mathbf{K}\mathbf{A})^{-1}\,\Delta\mathbf{P}' \tag{11.6}$$

Hence, it is necessary to consider only the change in loading as the new input and the formal state system model results.

As an illustration, consider the loading history for the compound bar of Figure 11.2 as given in Table 11.1. The specific system equations

Table 11.1

LOADING HISTORY FOR COMPOUND BAR OF FIG. 11.2

		LOADS IN LBS		RESULTING INCREMENTAL LOADING IN LBS	
	STATE	$\mathbf{P}'_B$	$\mathbf{P}'_C$	$\Delta\mathbf{P}'_B$	$\Delta\mathbf{P}'_C$
Initial Condition	$\mathbf{S}_0$	0	0	—	—
First Loading	$\mathbf{S}_1$	5	0	5	0
Second Loading	$\mathbf{S}_2$	5	8	0	8
Third Loading	$\mathbf{S}_3$	8	8	3	0

are:

$$\begin{bmatrix} u'_B \\ u'_C \end{bmatrix} = \begin{bmatrix} 0.11 & 0.08 \\ 0.08 & 0.24 \end{bmatrix} \begin{bmatrix} P'_B \\ P'_C \end{bmatrix} \tag{11.7}$$

Therefore,

$$\mathbf{S}_0 = \begin{bmatrix} u'_B \\ u'_C \end{bmatrix} = \begin{bmatrix} 0 \\ 0 \end{bmatrix}$$

$$\mathbf{P}'$$

$$\mathbf{S}_1 = \begin{bmatrix} 0.11 & 0.08 \\ 0.08 & 0.24 \end{bmatrix} \begin{bmatrix} 5 \\ 0 \end{bmatrix} = \begin{bmatrix} 0.55 \text{ in.} \\ 0.40 \text{ in.} \end{bmatrix}$$

$$\mathbf{S}_2 = \begin{bmatrix} 0.11 & 0.08 \\ 0.08 & 0.24 \end{bmatrix} \begin{bmatrix} 5 \\ 8 \end{bmatrix} = \begin{bmatrix} 1.19 \text{ in.} \\ 2.32 \text{ in.} \end{bmatrix} \tag{11.8}$$

$$\mathbf{S}_3 = \begin{bmatrix} 0.11 & 0.08 \\ 0.08 & 0.24 \end{bmatrix} \begin{bmatrix} 8 \\ 8 \end{bmatrix} = \begin{bmatrix} 1.52 \text{ in.} \\ 2.56 \text{ in.} \end{bmatrix}$$

and

$$\mathbf{\Delta P}'$$

$$\mathbf{S}_1 = \mathbf{S}_0 + \begin{bmatrix} 0.11 & 0.08 \\ 0.08 & 0.24 \end{bmatrix} \begin{bmatrix} 5 \\ 0 \end{bmatrix} = \begin{bmatrix} 0 \\ 0 \end{bmatrix} + \begin{bmatrix} 0.55 \\ 0.40 \end{bmatrix}$$

$$\mathbf{S}_2 = \mathbf{S}_1 + \begin{bmatrix} 0.11 & 0.08 \\ 0.08 & 0.24 \end{bmatrix} \begin{bmatrix} 0 \\ 8 \end{bmatrix} = \begin{bmatrix} 0.55 \\ 0.40 \end{bmatrix} + \begin{bmatrix} 0.64 \\ 1.92 \end{bmatrix} \tag{11.9}$$

$$\mathbf{S}_3 = \mathbf{S}_2 + \begin{bmatrix} 0.11 & 0.08 \\ 0.08 & 0.24 \end{bmatrix} \begin{bmatrix} 3 \\ 0 \end{bmatrix} = \begin{bmatrix} 1.19 \\ 2.32 \end{bmatrix} + \begin{bmatrix} 0.33 \\ 0.24 \end{bmatrix}$$

The state concept becomes extremely important if the properties of the structure as represented by $(\mathbf{A}^T\mathbf{K}\mathbf{A})^{-1}$ change as a result of loading the structure, as, for example, by strain hardening, buckling of members, or nonlinearity of material response. The concept is also useful when there are proposed modifications in the planned use of a structure. Suppose that after a structure has been designed, the owner wants to place heavier equipment than originally planned on an upper floor. Instead of repeating all his calculations, the designer can determine the response of the structure to the total loading as the sum of the response to the initial planned loads, which he has already calculated, and the response to the new load.

In general, Equation 11.9 can be rewritten as

$$\mathbf{S}_{i+1} = \mathbf{S}_i + (\mathbf{A}^T\mathbf{K}_i\mathbf{A})^{-1}\,\mathbf{\Delta P}_i \tag{11.10}$$

where $\mathbf{K}_i$ now represents the instantaneous stiffness characteristics of the structure as produced by the past loading and the current state, $\mathbf{S}_i$, of the structure.

Example 11.3   The simulation methods discussed in Chapter 8 provide another illustration of system states and state transformations.

The simulation model developed for the Ready Mix Plant provides an excellent example. In the example the simulation is built around the modeling and status of the parking lot, batching plant, and the receiving office for orders.

The parking lot status is related directly to the number of trucks it contains. Because the company owns five trucks, the parking lot can be in only one of six possible states; namely empty (S_0) or with one (S_1), two (S_2), three (S_3), four (S_4), or five (S_5) trucks parked. Figure 11.3

Number of trucks in parking lot	Parking lot state	Linear graph state model
0	S_0	⓪
1	S_1	①
2	S_2	②
3	S_3	③
4	S_4	④
5	S_5	⑤

Figure 11.3

Parking Lot States for Ready Mix Plant

illustrates these possible states and a linear graph state model.

The state transformations for the parking lot are simply related to the arrival or departure of trucks. Figure 11.4 shows the parking lot

Transformation effect of batching plant
calling for one truck

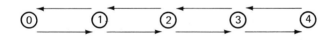

Transformation effect of single
truck return to parking lot

Figure 11.4

Parking Lot State Transformations for Order for One Truck for Ready Mix Plant

state transformations for single trucks and Figure 11.5 the transformations associated for a batching plant call for two trucks. Other state transformation models can be developed for batching plant requests for three, four, and five truck-full orders. In each case however, the parking lot state is equal to the number of trucks parked, and the transformations effect an addition or subtraction to the number of parked trucks.

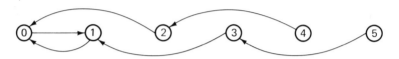

Figure 11.5

Parking Lot State Transformations for Order for Two Trucks for Ready Mix Plant

A state vector concept is also readily developed for the number of orders currently unfilled and held in the order receiving office. Again, the state transformation is related simply to the additions of an incoming order or the subtraction of an order released to the batching plant. Similarly, the batching plant can be empty awaiting a mix order, mixing, filling a truck, or full awaiting a truck.

A total state vector can be developed for the entire ready mix plant operation and corresponds to the aggregation of the systems component state vectors, as shown in Figure 11.6. The entire simulation process reduces to monitoring this system state vector at specific system times.

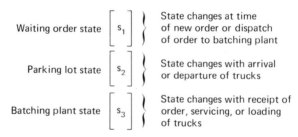

Figure 11.6

State Vector for Ready Mix Plant

In simulating many engineering problems, complex mathematical models and algorithms may be required to determine the system transformation and the time at which the transformation will occur. However, in all cases efforts are directed to determining the states of the system state vector. Consequently, simulation can be considered as a system state methodology.

11.3 DYNAMIC PROGRAMMING

Dynamic programming is an efficient enumeration procedure for determining the combination of decisions (changes in state variable values) that optimizes overall system effectiveness as measured by a criterion function. There exists no standard mathematical formulation for dynamic programming problems. It is a state theory approach, and the particular equations used must be developed to fit each individual problem. The dynamic programming approach to optimization is best explained by solving example problems.

Example 11.4   Consider the water supply system shown in Figure 11.7, which basically portrays the flow and use of water in a

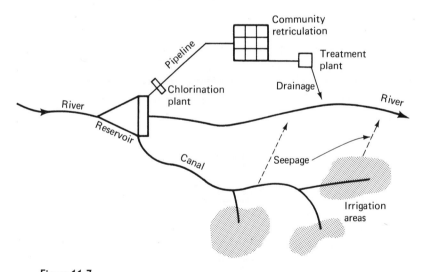

Figure 11.7

Typical City-Farmland Water Supply System

typical city-farmland situation. To describe the water distribution system in detail would require describing the system components and their interaction; i.e., river, dam, pipeline, chlorination plant, town retriculation systems, drainage system, waste treatment plant, irrigation canal, irrigation areas, seepage system flows, and so on. The individual component state vectors for each of these components would require many descriptions: physical, economic, social, and the like.

Suppose that the basic problem is concerned with water management and more specifically with the management of the water impounded by the reservoir; i.e., the owner wants to determine how much water to release each month in order to maximize his income from the sale of water. In this case, attempts might be made to model the reservoir in terms of the reservoir size; depth of water; surface area exposed to evaporation; streamflow characteristics; the community's annual, seasonal, and daily demands; agricultural demands; minimum streamflow requirements to maintain ecological, salinity, and boating conditions; and so on.

If, for example, too much water is released for irrigation, then future flows may not materialize, thus possibly jeopardizing the community's water supply. If, on the other hand, too little water is released, sudden future flows may cause the reservoir to fill and cause flooding. Clearly a management problem is involved in handling the storage volume.

The development of a major management control model for the reservoir requires, as a first step, the development of a suitable model for the reservoir storage volume. A simple model can be developed considering the reservoir purely from a storage volume point of view. For this model the volume of impounded water is the state variable, because the state of the reservoir can be described in terms of how full it is at any time, as shown in Figure 11.8. The state can change depending on the

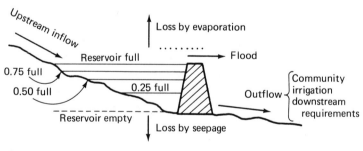

Figure 11.8

Water Reservoir System

nature of the inflow and outflow. Although inflow can easily be considered an environmental input to the system, the outflow, which is primarily the result of human action, can be considered a negative input from the storage volume point of view.

Because of its simplicity, a graphical concept will be used, as shown in Figure 11.9. Incremental volumes of quarter reservoir capacity

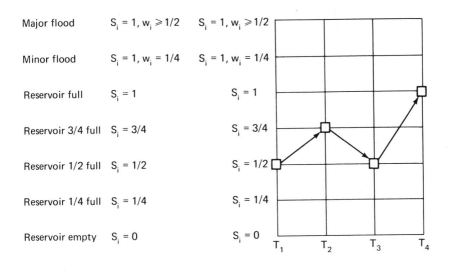

(a)   Potential System States   (b)   Typical System Behavior

Figure 11.9

Potential System States and Typical System Behavior for Reservoir System Model

are used to differentiate the various states of the reservoir; i.e., if the reservoir is half full at time T_i, then $S_i = 1/2$ where S_i represents the state of the reservoir. In Figure 11.9b a typical system response is shown in terms of the possible states and the system times T_1, T_2, T_3, and T_4. The actual changes in the system state are the result of both a management decision phase and an environment input phase, as shown in Figure 11.10. In Figure 11.10a the decision has been made to release water equivalent to a quarter reservoir volume, and in Figure 11.10b the local rainfall has produced a stream inflow equal to half the reservoir volume. Figure 11.10c models the interaction of the decisions to meet a commitment to supply water and the actual inflow due to nature. The actual seasonal inflow from the stream can be obtained from historical data and modeled as a chance node with probabilities associated with the various different inflows for each seasonal time increment. In spring, for

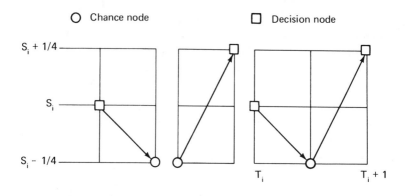

(a) Decision Phase   (b) Chance Phase   (c) System Time Increment

Figure 11.10

State Transformation for Reservoir System Model

example, there is a probability of P_1 that the inflow will be three quarter volume, P_2, that there will be half reservoir volume, and P_3 that there will be a quarter volume, as shown in Figure 11.11.

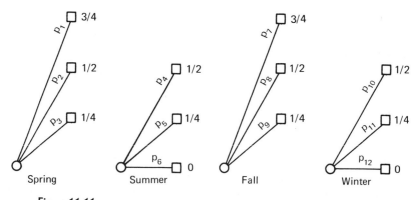

Figure 11.11

Seasonal Stream Flow

The basic state model for reservoir management is shown in Figure 11.12. Notice that at each decision node all possible decisions have been shown. Thus, if the reservoir is at the three-quarter level, four possible decisions can be made; namely, to release 0, 1, 2, or 3 quarter volumes of water. Also, for each chance node, all possible inflows have

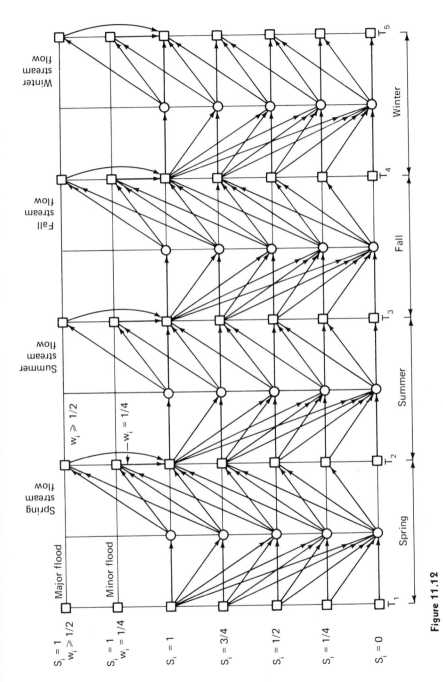

Figure 11.12

Reservoir System Management Model

been shown. If the reservoir is full (i.e., $S_i = 1$) and there is an inflow of half reservoir volume with no scheduled release, the succeeding state, S_{i+1}, is 1, which means that a major flood has resulted from a flood overflow and the reservoir is still full. A minor flood results from a flood overflow when the inflow plus amount in storage minus scheduled release exceeds the reservoir capacity by one-fourth of the reservoir volume.

The transformations involved in this state system model are simply scalar additions or subtractions such that

$$S_{i+1} = S_i - x_i + I_i - w_i \qquad (11.11)$$

where x_i is the quantity of water scheduled for release, I_i is the inflow, and w_i is the flood overflow during system time T_i to T_{i+1}. In general, however, the transformations correspond to the evaluation of, or the interpretation of, equations of state for the model and represent a major task.

The graphical model presented is similar in characteristics to those in Chapter 7 and similar solution techniques can be employed. The dynamic programming approach can be used very effectively with this problem. Before any of these solution techniques can be used, the model must be further developed to include the monetary values to be received from the sale of water to the city and farms, the monetary penalties that will be assessed if minor and major floods occur, the penalty cost to be assessed if the city water requirements are not met, and so on. These values may depend on the use of the water, how much is used, and when it is used.

Suppose that the owner of the reservoir limits his problem to determining how much water to release from the reservoir in each of the four time periods: spring, summer, fall, and winter. Suppose that the expected stream flow for the spring, summer, fall, and winter are known and are one-half, one-fourth, one-half, and one-fourth of the reservoir capacity, respectively. For convenience, the reservoir capacity is assumed to be 1 unit of water; i.e., when the reservoir is full it contains 1 unit of water. It is also assumed that the releases to be made will be 0 or $\frac{1}{4}$ multiples of the reservoir capacity. This assumption is for convenience only, because releases can be of any volume. However, the computational effort required to solve the problem increases as the number of possible release conditions increases. During any time period, \$2,000 will be received for the first quarter unit of water released, \$1,500 for the second quarter unit, \$1,000 for the third quarter unit, and \$500 for each additional quarter unit of released water. Therefore, the return or income, as a function of amount of water released, is given by Table 11.2. If a minor flood occurs during any period ($w_i = \frac{1}{4}$ unit), a \$1,500 penalty must be paid; and if a major flood occurs, ($w_i = \frac{1}{2}$ unit), a

$4,000 penalty must be paid and the flood water will be wasted; i.e., the flood water cannot be sold. Furthermore, suppose that the reservoir is three-quarters full at the beginning of the spring period.

Table 11.2

RETURNS AS A FUNCTION OF RESERVOIR RELEASES

AMOUNT RELEASED, x_i, IN UNITS OF RESERVOIR CAPACITY	TOTAL RETURN $ $r_i(x_i, w_i)$		
	If No Flood $w_i = 0$	If Minor Flood $w_i = \frac{1}{4}$	If Major Flood $w_i = \frac{1}{2}$
0	0	−1,500	−4,000
1/4	2,000	500	−2,000
1/2	3,500	2,000	− 500
3/4	4,500	3,000	500
1	5,000	3,500	1,000
$1\frac{1}{4}$	5,500	4,000	1,500

In order to maximize returns from selling water, the criterion function to be maximized is

$$Z = r_1(x_1, w_1) + r_2(x_2, w_2) + r_3(x_3, w_3) + r_4(x_4, w_4) \qquad (11.12)$$

where $r_i(x_i, w_i)$ represents the return for the ith period due to the scheduled release x_i and the flood overflow w_i for the ith period. The value of $r_i(x_i, w_i)$ can be obtained from Table 11.2. This equation is to be maximized subject to the condition that no more water can be sold in any period than there is water available during that period. Therefore,

$$x_i \leq S_i + I_i \qquad (11.13)$$

where S_i represents the amount of water in the reservoir at the time period T_i, and x_i and I_i represent the amount of release and inflow, respectively, for the time period from T_i to T_{i+1}, (i.e., x_1 represents the amount of water scheduled for release during the spring beginning at time T_1). Furthermore,

$$x_i \geq 0 \qquad (11.14)$$

because a negative quantity of water cannot be sold. The state of the reservoir at time T_{i+1} can be computed from the transformation equation given by Equation 11.11 and is

$$S_{i+1} = S_i - x_i + I_i - w_i \qquad (11.15)$$

where S_{i+1} must be less than or equal to S_{max} and greater than or equal to S_{min}. In this case S_{max} is 1 and S_{min} is 0.

Dynamic programming can be used to solve this problem if the criterion function can be written in the form of a recurrence equation. To develop a formal recurrence equation for this problem, let $f_n(S_n)$ be a cumulative return function at time T_n such that

$$f_5(S_5) = \max \left[r_1(x_1, w_1) + r_2(x_2, w_2) + r_3(x_3, w_3) + r_4(x_4, w_4) \right] \quad (11.16)$$

where $f_5(S_5)$ represents the maximum accumulated return if the system state at time T_5 is given by S_5. At time T_2

$$f_2(S_2) = \max_{x_1} r_1(x_1, w_1) \quad (11.17)$$

subject to:

$$S_2 = S_1 - x_1 + I_1 - w_1 \quad (11.18)$$

$$0 \le S_2 \le 1 \quad (11.19)$$

$$S_1 = \tfrac{3}{4} \quad (11.20)$$

$$x_1 \le S_1 + I_1 \quad (11.21)$$

$$w_1 \text{ and } x_1 \ge 0 \quad (11.22)$$

where max means that the following expression is to be maximized by x_i
varying x_i.

Equation 11.18 is the transformation equation and defines the state in which the reservoir will be at T_2 if the reservoir state at T_1 is S_1, x_1 units are released, I_1 units of inflow are received and w_1 units are lost as flood overflow. Equation 11.19 limits the state of the reservoir at time T_2 to a state somewhere between empty and full. If w_1 must be greater than 0 in order to satisfy Equation 11.19, a flood occurs and the value for $r_1(x_1, w_1)$ is determined from column 3 or 4 of Table 11.2 depending on whether w_1 is $\tfrac{1}{4}$ or $\tfrac{1}{2}$ unit. Equation 11.20 specifies the initial state of the reservoir at time T_1. Equation 11.21 specifies the upper limit on what the releases can be, and Equation 11.22 specifies that releases and flood overflows cannot be negative.

If $x_1 = 0$, then at time T_2, $S_2 = \tfrac{3}{4} - 0 + \tfrac{1}{2} - w_1 = \tfrac{5}{4} - w_1$, which requires that $w_1 = \tfrac{1}{4}$ in order to satisfy Equation 11.19. Thus, a minor flood occurs that results in a value of $-\$1,500$ for $r_1(0, \tfrac{1}{4})$ and $S_2 = 1$. Hence, $f_2(1) = -\$1,500$ as a first trial. If $x_1 = \tfrac{1}{4}$, then $S_2 = \tfrac{3}{4} - \tfrac{1}{4} + \tfrac{1}{2} - w_1 = 1 - w_1$, such that $w_1 = 0$ and $r_1(\tfrac{1}{4}, 0) = \$2,000$. This value is larger than the other value for $S_2 = 1$. Therefore, $f_2(1)$ becomes $\$2,000$. If $x_1 = \tfrac{1}{2}$, then $S_2 = \tfrac{3}{4} - \tfrac{1}{2} + \tfrac{1}{2} - w_1 = \tfrac{3}{4} - w_1$ such that $w_1 = 0$ and $r_1(\tfrac{1}{2}, 0) = \$3,500$. Therefore, $f_2(\tfrac{3}{4}) = \$3,500$. Similarly, for $x_1 = \tfrac{3}{4}$,

$f_2(\frac{1}{2}) = \$4,500$; for $x_1 = 1$, $f_2(\frac{1}{4}) = \$5,000$; and for $x_1 = \frac{5}{4}$, $f_2(0) = \$5,500$. The results for time T_2 are

$$f_2(1) = \$2,000$$
$$f_2(\tfrac{3}{4}) = \$3,500$$
$$f_2(\tfrac{1}{2}) = \$4,500 \qquad (11.23)$$
$$f_2(\tfrac{1}{4}) = \$5,000$$
$$f_2(0) = \$5,500$$

The maximum return that can be obtained for every possible state that the reservoir can be in at time T_2 and the release that must be made to achieve this return has now been computed. The result is shown in Figure 11.13. However, the state at time T_2 that will lead to the

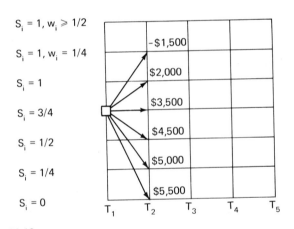

Figure 11.13

Maximum Returns for Each State at T_2 for Reservoir System Model

maximum returns for the entire four periods cannot yet be determined, because the state at T_2 will affect the amount of water that can be released during the next period, and therefore affects future returns.

The next step is to calculate the maximum total returns that can be obtained for every state at time T_3. Now at time T_3, the cumulative return function at each possible state is equal to the maximum return that can be obtained for all possible combinations of states at T_2 and releases that will allow the state at T_3 to be attained. Therefore,

$$f_3(S_3) = \max_{x_2} [r_2(x_2, w_2) + f_2(S_2)] \qquad (11.24)$$

subject to

$$S_3 = S_2 - x_2 + I_2 - w_2 \qquad (11.25)$$

$$0 \leq x_2 \leq S_2 + I_2 \qquad (11.26)$$

$$0 \leq S_2 \leq 1 \qquad (11.27)$$

$$0 \leq S_3 \leq 1 \qquad (11.28)$$

$$w_2 \geq 0 \qquad (11.29)$$

Equation 11.25 is the transformation equation that determines the value of S_3 that can be attained for different combinations of S_2 and x_2. From Equation 11.25 there are three combinations of S_2 and x_2 that will result in $S_3 = 1$ for $I_2 = \frac{1}{4}$. If $S_2 = 1$ and $x_2 = 0$, $S_3 = 1$ when $w_2 = \frac{1}{4}$, which means that a minor flood occurs and $\frac{1}{4}$ unit of water must be wasted. If $S_2 = 1$ and $x_2 = \frac{1}{4}$, $S_3 = 1$ and $w_2 = 0$; and if $S_2 = \frac{3}{4}$ and $x_2 = 0$, then $S_3 = 1$ and $w_2 = 0$. Therefore,

$$f_3(1) = \max \{[r_2(0, \tfrac{1}{4}) + f_2(1)] \text{ or } [r_2(\tfrac{1}{4}, 0)$$
$$+ f_2(1)] \text{ or } [r_2(0, 0) + f_2(\tfrac{3}{4})]\}$$

$$f_3(1) = \max \{[-1,500 + 2,000] \text{ or } [2,000$$
$$+ 2,000] \text{ or } [0 + 3,500]\}$$

$$f_3(1) = \$4,000 \text{ for } x_2 = \tfrac{1}{4} \text{ and } S_2 = 1$$

Similarly,

$$f_3(\tfrac{3}{4}) = \max \{[r_2(\tfrac{1}{2}, 0) + f_2(1)] \text{ or } [r_2(\tfrac{1}{4}, 0)$$
$$+ f_2(\tfrac{3}{4})] \text{ or } [r_2(0, 0) + f_2(\tfrac{1}{2})]\}$$

$$f_3(\tfrac{3}{4}) = \max \{[3,500 + 2,000] \text{ or } [2,000 + 3,500] \text{ or } [0 + 4,500]\}$$

$$f_3(\tfrac{3}{4}) = \$5,500 \text{ for } x_2 = 0 \text{ and } S_2 = 1 \text{ or } x_2 = \tfrac{1}{4} \text{ and } S_2 = \tfrac{3}{4}$$

This means that there are two values of x_2 that provide the same possible maximum accumulated return at $S_3 = \frac{3}{4}$. Direct calculations lead to $f_3(\frac{1}{2}) = \$7,000$, $f_3(\frac{1}{4}) = \$8,000$, and $f_3(0) = \$9,000$. The results for time T_3 are shown in Figure 11.14.

At time T_4, the cumulative return becomes

$$f_4(S_4) = \max [r_3(x_3, w_3) + f_3(S_3)] \qquad (11.30)$$

subject to

$$S_4 = S_3 - x_3 + I_3 - w_3 \qquad (11.31)$$

$$0 \leq x_3 \leq S_3 + I_3 \qquad (11.32)$$

$$0 \leq S_3 \leq 1 \qquad (11.33)$$

$$0 \leq S_4 \leq 1 \qquad (11.34)$$

$$w_3 \geq 0 \qquad (11.35)$$

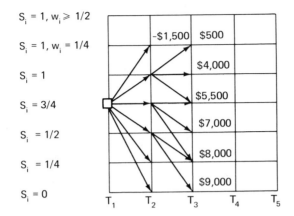

Figure 11.14

Maximum Returns for Each State at T_3 for Reservoir System Model

Notice now that Equations 11.30 through 11.35 are different from Equations 11.24 through 11.29 only in that subscripts 3 and 2 in Equations 11.24 through 11.29 have been replaced with 4 and 3, respectively, in Equations 11.30 through 11.35. Therefore, the general cumulative return can be written as

$$f_{i+1}(S_{i+1}) = \max_{x_i} [r_i(x_i, w_i) + f_i(S_i)] \qquad i = 1, 2, 3, 4 \quad (11.36)$$

subject to

$$S_{i+1} = S_i - x_i + I_i - w_i \tag{11.37}$$

$$0 \leq x_i \leq S_i + I_i \tag{11.38}$$

$$S_{\min} \leq S_i \leq S_{\max} \tag{11.39}$$

$$S_{\min} \leq S_{i+1} \leq S_{\max} \tag{11.40}$$

$$w_i \geq 0 \tag{11.41}$$

where $S_{\min}$ and $S_{\max}$ are the minimum and maximum values, respectively, that the reservoir state can assume and $f_1(S_1)$ is zero because it is the value of the cumulative return before any water is sold.

Equation 11.36 is known as the recurrence equation of dynamic programming. The student should solve this equation for times T_4 and T_5 to obtain the results for the water management example that are indicated in Figure 11.15.

Only after the computations for the last time period have been made can the maximum return for the entire four periods be determined. In this problem the maximum return is $15,000 when the reservoir is empty at T_5. If there had been a constraint such that the reservoir must

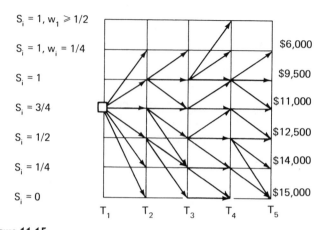

Figure 11.15

Maximum Returns for Each State at T_5 for Reservoir System Model

always have at least $\frac{1}{4}$ unit of water in it, then the maximum return would have been \$14,000.

The optimal set of states can be determined by finding a path in Figure 11.15 from the initial state of the reservoir at $S_1 = \frac{3}{4}$ to the state $S_5 = 0$ for which the optimal value of the returns occur. This can be done by starting at state $S_5 = 0$ and tracing a chain back to the initial state $S_1 = \frac{3}{4}$ such that the chain is composed of only as many branches as there are time periods. Thus, an optimal set of states is $S_1 = \frac{3}{4}$, $S_2 = \frac{3}{4}$, $S_3 = \frac{1}{2}$, $S_4 = \frac{1}{2}$, and $S_5 = 0$. The optimal series of releases can then be obtained by rewriting Equation 11.37 to solve for x_i. Hence

$$x_1 = S_1 - S_2 + I_1 - w_1 = \tfrac{3}{4} - \tfrac{3}{4} + \tfrac{1}{2} - 0 = \tfrac{1}{2}$$
$$x_2 = S_2 - S_3 + I_2 - w_2 = \tfrac{3}{4} - \tfrac{1}{2} + \tfrac{1}{4} - 0 = \tfrac{1}{2}$$
$$x_3 = S_3 - S_4 + I_3 - w_3 = \tfrac{1}{2} - \tfrac{1}{2} + \tfrac{1}{2} - 0 = \tfrac{1}{2}$$
$$x_4 = S_4 - S_5 + I_4 - w_4 = \tfrac{1}{2} - 0 + \tfrac{1}{4} - 0 = \tfrac{3}{4}$$

This set of optimal values of x_i is known as an optimal policy. There are three other sets of values for x_i that will also produce the optimal value of \$15,000 for the cumulative return. The student should find these.

The only conditions necessary for decomposing problems so that the recurrence equation can be developed are that the criterion function be separable and monotonic (Nemhauser, 1966). The criterion function is separable if the return function $r_i(x_i)$ for each x_i can be defined such that it is independent of the other variables x_j, $j \neq i$. The criterion function is monotonic if $r_i(x_i + \epsilon) \geq r_i(x_i)$ where ϵ is a small positive number.

In dynamic programming, constraints can be a help or a hindrance, depending on the form of the problem. In many cases, they limit the range of the x_i variables and therefore reduce the computational complexity. However, if a constraint relates two or more x_i variables, then it is necessary to add another state variable and this greatly increases the computational complexity because the recurrence equation then becomes

$$f_{i+1}(S_{1,i+1}, S_{2,i+1}) = \max_{x_i} \left[r_i(x_i) + f_i(S_{1,i}, S_{2,i}) \right]$$

Let each state variable have L discrete values. Previously, the recurrence equation only had to be solved L times at each time T_i. Now it has to be solved L^2 times; one for each combination of $S_{1,i}$ and $S_{2,i}$. In general, the number of computations increases exponentially as the number of state variables.

Dynamic programming has most frequently been applied to problems that can be described in terms of several stages over time, with each stage representing a time period. However, dynamic programming itself has nothing to do with time. A stage is a point, either in space or time, at which a decision is required and may have no connotation of time.

Example 11.5 Assume that the irrigation region shown in Figure 11.7 has 3 units of water available to it and that the operator must determine how much water each of the 3 areas is to receive. This area can be represented by the pipe distribution system shown in Figure 11.16.

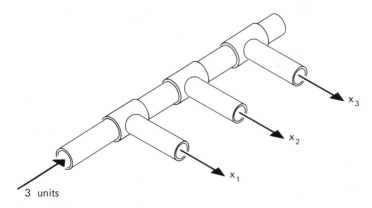

3 units

Figure 11.16

Pipe Distribution System Supplying Three Separate Outlets

The returns for each area as a function of the water released to that area are given in Table 11.3. The operator wants to allocate the water such

Table 11.3

RETURN FUNCTIONS FOR PIPE SYSTEMS

AMOUNT OF WATER RELEASED AT OUTLET i x_i (Units)	RETURNS		
	Outlet 1 $r_1(x_1)$ $	Outlet 2 $r_2(x_2)$ $	Outlet 3 $r_3(x_3)$ $
1	3	1	4
2	6	3	5
3	8	9	6

that returns from the 3 units of water are maximized. Therefore, the criterion function to be maximized is

$$Z = r_1(x_1) + r_2(x_2) + r_3(x_3)$$

subject to:

$$x_1 + x_2 + x_3 \le 3$$

$$x_i \ge 0$$

The solution procedure for this problem is as follows:

Step 1. Determine what will be represented by stages, state variables, and decision variables. Here a stage represents a pipe outlet, the state variable represents the amount of water available for that and all previous stages, and the decision variable represents the amount of water released at each stage.

Step 2. Convert the criterion function to a recurrence equation and constraints such that

$$f_i(S_i) = \max_{x_i} [r_i(x_i) + f_{i-1}(S_{i-1})] \qquad i = 1, 2, 3$$

$$S_i \le 3; \qquad S_{i-1} = S_i - x_i; \qquad x_i \le S_i; \qquad x_i \ge 0; \qquad f_0(S_0) = 0$$

Step 3. Compute $f_1(S_1)$ for $S_1 = 1, 2,$ and 3 units of water available at stage 1. This is equal to the returns for outlet 1 in Table 11.3.

Step 4. For $S_2 = 1, 2,$ and 3, compute the maximum $f_2(S_2)$ for each possible combination of x_2 and $S_2 - x_2$. This is summarized as stage 2 results in Table 11.4 where the maximum value of $f_2(S_2)$ for each S_2 is circled.

Step 5. For $S_3 = 1, 2,$ and 3, compute the maximum $f_3(S_3)$ for each possible combination of x_3 and $S_3 - x_3$ using the $f_2(S_3 - x_3)$ computed in step 4. This is summarized as stage 3 results in Table 11.5 where the maximum values of $f_3(S_3)$ for each S_3 is circled.

Table 11.4

STAGE 2 RESULTS FOR DYNAMIC PROGRAMMING APPROACH TO WATER DISTRIBUTION PROBLEM

UNITS OF WATER AVAILABLE TO STAGES 1 AND 2 S_2	UNITS OF WATER RELEASED AT STAGE 2 x_2	UNITS OF WATER RELEASED AT STAGE 1 $(S_2 - x_2)$ S_1	\$ RETURNED AT STAGE 2 (FROM TABLE 11.3) $r_2(x_2)$	\$ RETURNED FOR STAGE 1 $f_1(S_1)$ $r_1(x_1)$ (FROM TABLE 11.3)	TOTAL RETURN \$ $f_2(S_2)$
1	1	0	1	0	1
	0	1	0	3	③
2	2	0	3	0	3
	1	1	1	3	4
	0	2	0	6	⑥
3	3	0	9	0	⑨
	2	1	3	3	6
	1	2	1	6	7
	0	3	0	8	8

Table 11.5

STAGE 3 RESULTS FOR DYNAMIC PROGRAMMING APPROACH TO WATER DISTRIBUTION PROBLEM

UNITS OF WATER AVAILABLE TO STAGES 1, 2, 3 S_3	UNITS OF WATER RELEASED AT STAGE 3 x_3	UNITS OF WATER AVAILABLE TO STAGES 1 AND 2 S_2	\$ RETURNED AT STAGE 3 (FROM TABLE 11.3) $r_3(x_3)$	\$ RETURNED FROM STAGES 1 AND 2 (FROM STAGE 2 RESULTS) $f_2(S_2)$	TOTAL RETURN \$ $f_3(S_3)$
1	1	0	4	0	④
	0	1	0	3	3
2	2	0	5	0	5
	1	1	4	3	⑦
	0	2	0	6	6
3	3	0	6	0	6
	2	1	5	3	8
	1	2	4	6	⑩ *
	0	3	0	9	9

Step 6. Choose the maximum value of $f_3(S_3)$. This value is starred in stage 3 results.

Step 7. Now go back through the problem and determine the optimal values of the x_i. From stage 3 results, the maximum $f_3(S_3)$ is 10 for $x_3 = 1$ with 2 units divided between stages 1 and 2; i.e., $S_2 = S_3 - x_3 = 3 - 1 = 2$. From stage 2 results, the optimum use of 2 units; i.e., $S_2 = 2$, results in an optimum return of 6 where $x_2 = 0$ and $x_1 = 2$ units. Therefore, the solution is $x_1 = 2$, $x_2 = 0$, and $x_3 = 1$ for a maximum return of \$10.

Example 11.6   A decision tree is shown in Figure 11.17. If the tree is flipped to the position shown in Figure 11.18, the problem can be

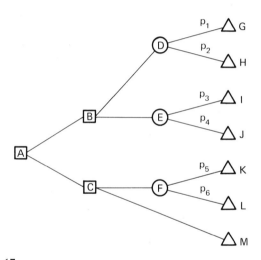

Figure 11.17

Decision Tree

formulated as a dynamic programming problem where a state represents a consequent or decision node and stage 1 represents the set of consequent states G through M, stage 2 represents the set of decision states B and C, and stage 3 represents the set of decision states A. The transformation is the decision that leads to a new decision node, and the criterion is to maximize the expected return where $f_1(S_1)$ for $S_1 = G, H, \ldots, N$ is equal to the value at the consequent node. Thus,

$$f_2(B) = \max \{[P_1 f_1(G) + P_2 f_1(H)] \text{ or } [P_3 f_1(I) + P_4 f_1(J)]\}$$
$$f_2(C) = \max \{[P_5 f_1(K) + P_6 f_1(L)] \text{ or } [f_1(M)]\}$$

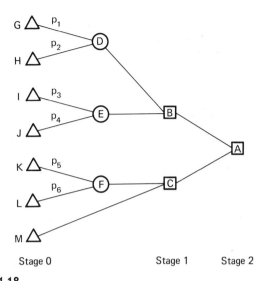

Stage 0   Stage 1   Stage 2

Figure 11.18

Flipped Decision Tree

and

$$f_3(A) = \max \{[f_2(B)] \text{ or } [f_2(C)]\}$$

Example 11.7 Recall the traffic flow problem of Section 2.4 where the objective was to find the maximum number of vehicles per hour that can go from point 1 to point 5. The dual of this problem (Figure 2.9) can be formulated as a dynamic programming problem in which the states represent the nodes A, B, C, D, and E and stage 1 represents the set of states A, stage 2 represents the set of states B, stage 3 represents the set of states C and D, and stage 4 represents the set of states E. The criterion is the minimum path through the dual and the transformation is the decision about which state to come from to get to a given state. Thus,

$$f_1(A) = 0$$

$$f_2(B) = \text{Min } \{[700 + f_1(A)] \text{ or } [800 + f_1(A)]\} = 700$$

$$f_3(C) = \text{Min } \{[900 + f_2(B)] \text{ or } [400 + 1,000 + f_2(B)]\}$$

$$f_3(C) = \text{Min } \{[1,600] \text{ or } [2,100]\} = 1,600$$

$$f_3(D) = \text{Min } \{[900 + 400 + f_2(B)] \text{ or } [1,000 + f_2(B)]\}$$

$$f_3(D) = \text{Min } \{[2,000] \text{ or } [1,700]\} = 1,200$$

$$f_4(E) = \text{Min } \{[1,000 + f_3(C)] \text{ or } [700 + f_3(D)]\}$$

$$f_4(E) = \text{Min } \{[2,600] \text{ or } [2,400]\} = 2,400$$

Therefore, the optimum value is for the path composed of branches 1, 4, and 7, which correspond to roads 1-3, 4-5, and 7-5 and is the same as determined previously.

11.4 SUMMARY

State concepts can be used to study system behavior. The system status at any time is uniquely and adequately defined or described by the state variables. The transformation function defines the new state that results from a change in the value of a state variable. Simulation is an analysis procedure in which the efforts are directed to determining the value of the system state vector at different times.

If a measure of utility can be assigned to system states or to changes in state variables, or to both, the decision-maker may want to optimize the utility that can be obtained from the system. Dynamic programming is an optimization procedure that is based on state theory concepts.

Although the application of the dynamic programming approach depends on the ingenuity of the problem-solver, there are certain characteristics that a problem must have. These characteristics are:

1. The problem must be one that can be divided into stages with a decision required at each stage. The stages may represent different points in space, as, for example, in selecting a route for a new pipeline, or they may represent different points in time, as in determining the optimal releases from a reservoir, or they may represent different activities, as in determining how much of available resources to allocate to different uses.

2. Each stage of the problem must have a finite number of states associated with it. The states describe the possible conditions in which the system might be at that stage of the problem. In the reservoir problem, the states represented the amount of water stored in the reservoir at that stage. In the distribution problem, the states represented the amount of water to be distributed to that and all preceding stages.

3. The effect of a decision at each stage of the problem is to transform the current state of the system into a state associated with the next stage. The decision may represent how much water to release from the reservoir at the current time, and this decision will transform the amount of water stored in the reservoir from the current amount to a new amount for the next stage.

4. Associated with each potential state transformation is a return that indicates the effectiveness or utility of the transformation.

This means that there must be a return function defined for each stage.

5. The optimality of future decisions can be judged on the basis of the current state of the system.

Dynamic programming is a simple procedure from the computational point of view, and one that can treat nonconvex, nonlinear, discontinuous criterion, and constraint functions. Extensive examples of applications are presented by Bellman (1957), Bellman and Dreyfus (1962), Wagner (1969), and Hall and Dracup (1970). The limiting factor in dynamic programming has been the fact that it becomes inefficient if more than two state variables are present. However, recent studies have shown that with certain modifications, dynamic programming can be used to efficiently solve problems that contain four state variables (Heidari *et al.*, 1971).

11.5 PROBLEMS

P11.1. Systems can be described by state vectors in terms of the values adopted by state variables. Different descriptive state vectors can be developed depending on the complexity of the system and the purpose of the description. Identify a number of compatible purposes and state vectors for the areas listed below. How would you establish the size of the descriptive vector and the order of importance of the state variables.

1. A moving pendulum
2. A drawbridge
3. A storm water discharge pipe with flap valve
4. A house telephone
5. A highway toll collection area
6. A shipping harbor
7. A railroad switching yard

P11.2. Formulate the blending problem of the Sunnyflush Company in Section 5.3 as a state theory problem.

1. What are the stages, state variables, and decision variables?
2. Develop the recurrence equation, the transformation function, and the constraints for a dynamic programming model of the problem.
3. Solve the dynamic programming problem.

P11.3. Figure P11.3 illustrates the conditions at Washington and Anderson streets. Left turns are not permitted at this intersection. You have been assigned to develop a state theory model for the auto traffic going east on Washington at this intersection.

1. Why would the time that the light is green not be a good state variable?

2. If the number of cars at the light waiting to cross is chosen as the state variable, write the state equation; be sure to define terms in equations.

3. How would you obtain data for the terms in your state equation?

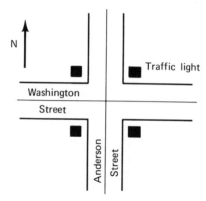

Figure P11.3

P11.4. You currently have a 2-year-old machine in your plant. The machine deteriorates with age such that the annual operating expenses increase each year. This type of machine will be needed in your plant for the next 10 years. The cost of buying a machine, its trade-in value, and expected operating expenses for 1 year are a function of the machine's age and are given by:

AGE IN YEARS	COST	TRADE-IN VALUE	OPERATING EXPENSE
i	C_i	T_i	E_i
0	$4000	$3200	$100
1	2000	1700	150
2	1000	700	220
3	700	400	350
4	500	300	450
5	350	200	550

Thus a used machine 2 years old would cost $1,000 to purchase and $220 to operate for 1 year, but is worth only $700 as a trade in on another machine. At what dates should you trade in the machine and how old a machine should you buy each time to minimize the total cost of the machine for the next 10 years? Formulate the problem as a state theory problem and solve by dynamic programming.

P11.5. Problem 8.5 is concerned with the simulation of an elevator operation in a high-rise building.

1. How can this problem be formulated using state theory concepts? What are the possible states and transformations?

2. If three elevators exist, how would you use state theory concepts in the analysis of the problems? Can you incorporate in your approach the formulation given in your answer to (1) above?

P11.6. Solve the reservoir problem in Section 11.4 if:

1. The expected streamflows for spring, summer, fall, and winter, are $\frac{1}{2}$, 1, $\frac{1}{2}$, and $\frac{3}{4}$ of the reservoir capacity, respectively.

2. The probabilities for the seasonal streamflow as shown in Figure 11.7 are:

SPRING	SUMMER	FALL	WINTER
$p_1 = 0.2$	$p_4 = 0.1$	$p_7 = 0.2$	$p_{10} = 0.3$
$p_2 = 0.6$	$p_5 = 0.8$	$p_8 = 0.7$	$p_{11} = 0.4$
$p_3 = 0.2$	$p_6 = 0.1$	$p_9 = 0.1$	$p_{12} = 0.3$

(Hint: When formulating dynamic programming problems that involve probabilities, it is often convenient to use the same solution approach as is used in solving decision tree problems by dynamic programming.)

P11.7. Formulate the production scheduling problem of the Hard Rock Company in Section 5.3 as a state theory problem and solve by dynamic programming. How would the results differ if the plant could not work any overtime?

P11.8. Many problems faced by engineers are so complex that there is no readily apparent equation of state or they contain so many state variables that all the possible states cannot be defined.

All proposed major engineering works must be accompanied by an environmental impact statement that delineates the effect that the project will have on the environment. Since the current state of the art concepts and modeling of the environment are limited, many problems are raised concerning the question of which state variables are to be included in the state vector used to describe the environment and how to define the possible transformations that occur as the result of the project implementation.

What are some of the state variables you would want to include in the state vector, and how would you approach the problem of developing the transformations required to determine the effect of each of the following proposed projects on the environment?

1. A highway
2. A manufacturing plant
3. A high-rise building
4. A reservoir
5. A construction project
6. A bridge across a rural stream

12

SYSTEMS CONCEPTS
IN ENGINEERING

Aerial view of the A. D. Edmonston Pumping Plant, part of the California State Water Project, the largest single water development in the world. The California State Water Project was selected as the Outstanding Civil Engineering Achievement of 1972. The project will provide throughout its 600 mile route a firm supply of good quality water, a guaranteed source of clear hydroelectric power, flood control, recreation sites readily available to urban areas, and the enhancement of fish and wildlife habitats through wise water management and conservation. (Courtesy of the American Society of Civil Engineers)

12.1   ENGINEERING RESPONSE TO SOCIETAL NEEDS

In practice, the engineer often encounters a societal need in the form of a problem. Either he brings the problem to the attention of his client, or the client presents the engineer with the problem. The engineer should distinguish between the true problem and the symptoms of a problem. For example, traffic congestion is really a symptom of a traffic problem and not the actual cause of a problem. Traffic congestion can be the result of any number of things, such as inadequate road design, insufficient zoning controls, or lack of alternate forms of transportation.

Once the engineer has identified the problem, he must determine at what level in the hierarchical structure of the problem he will approach it and what goals and objectives are to be achieved by solving the problem. The level at which the problem is approached and the goals and objectives to be achieved are interrelated.

Having determined the goals and objectives and the level at which he will approach the problem, the engineer must propose a system solution that will best fulfill the goals and objectives. This proposed system will be composed of components and, therefore, he will be concerned with establishing the behavior and gaining an understanding of the components of the system. He must then evaluate how well the proposed solution performs in view of the stated goals and objectives. He will also be concerned with implementing the solution. A proposed solution is of no use unless it can be implemented.

At each stage of his response, the engineer will be using data and models that he already has or that he must acquire or develop. Sometimes a model will require data that must be obtained by observation and measurement. At other times, the available data may limit the types of models that can be developed and used.

This entire process may be performed by one engineer or it may require a project team to complete the study and evaluation, depending on the size and complexity of the system under consideration.

This chapter will attempt to demonstrate how problems might be approached using some of the system concepts already discussed. Although not representing any actual set of conditions and solutions, the following illustrates how a problem might be approached at different levels and the kinds of solutions that might be expected from these approaches.

12.2   PROBLEMS

A number of major, medium, and minor towns are distributed over a section of the country, as shown in Figure 12.1. Town A has been sub-

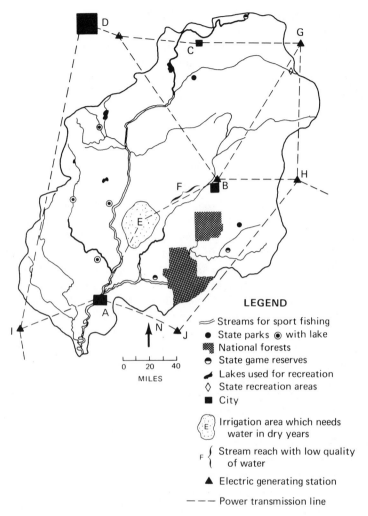

LEGEND

≈ Streams for sport fishing
● State parks ◉ with lake
▨ National forests
◓ State game reserves
◤ Lakes used for recreation
◊ State recreation areas
■ City

(E) Irrigation area which needs
    water in dry years

F { Stream reach with low quality
    of water

▲ Electric generating station

— — — Power transmission line

0   20   40
MILES

Figure 12.1

Region with Problems

jected to extensive flooding the past few years and the city council has asked the government for aid in combating the problem. Town B has been experiencing a water shortage during the summer months of dry years. The farmers in the area denoted by E have combined to form a cooperative to obtain water for irrigation during dry years. The regional conservation club has begun a campaign to inform the population that the quality of water in reach F of the stream has deteriorated such that fish can no longer survive there.

These problems have been brought to the attention of the state department of planning and economic development. They may not even be the most critical problems in the region; however, they are the ones that have received attention because of the efforts of those affected to have these problems solved.

12.3 REGIONAL APPROACH TO PROBLEMS

The problems stated above are symptoms that indicate that the people of the region have not been able to attain their goals. Before any solution is proposed for these problems, the goals and objectives should be expressed explicitly. The main goal in this case may be the maximization of regional welfare, where the term "region" can be employed to denote a geographic area ranging from the size of a small farming field to a large nation. Objectives that will contribute to the attainment of this goal may be to generate an increase in income to individuals in the region, develop facilities that provide benefits in excess of their costs, promote and support economic growth, or defend or preserve natural environmental conditions.

If the problems are approached on an individual basis, the proposed solutions may increase the number and severity of the problems rather than solve them. For example, one alternative for reducing the flooding at A is to build a flood control reservoir upstream from the town to store flood waters for release during months when flooding does not occur. One alternative for increasing the water supply for Town B might be a water supply reservoir upstream from Town B. If these problems are thus approached as individual problems, two reservoirs may be required. However, if they are considered simultaneously, then one reservoir might be able to satisfy both the water supply requirements at B and the flood control requirements at A for a much lower cost than two reservoirs.

Therefore, the larger the region considered, the more likely that all factors of influence will be within the region being considered. However, the complexity and difficulty of the problem increases with increasing size of the region considered. Thus, the region chosen for study, which is shown in Figure 12.1, may be a compromise between including all the influencing factors and keeping the problem small enough to be solved within time, budget, and manpower resources available.

Of the many possible alternatives for achieving the objectives stated above, one is a water resources system along with waste treatment plants to reduce the waste discharge into the stream that results in low water quality. If irrigation water is supplied, it may require extensive pumping and this may increase the power requirements above the present supply capabilities. Thus, power supply may become an additional purpose for the development. The points of power demand and supply may be connected in a power network, as shown in Figure 12.1. Hence the influence of power supply and demand from outside the region must be incorporated into any model constructed.

Next, criteria must be developed that can be used to measure how well various proposed alternatives satisfy the stated objectives. One criterion that might be applicable is dollars. The flood damage reduction might be measured in dollars. Power, which can be sold, may also be measured in dollar terms. However, the effects of pollution may be difficult to evaluate in dollar terms. Often when commensurate criteria cannot be developed, the benefits for one of the purposes may be expressed as a standard; i.e., as a level that must be provided. Hence, this standard becomes a constraint that must be met.

The next step is to identify as many alternatives as possible that might satisfy the stated objectives. At this step, the main concern is identifying possible alternatives, not the optimal or best alternatives. These possible alternatives should include not only those possible under existing technology, but also those possible under technology expected to exist at the time the alternative is needed in the system. The important point is that the only alternatives that will be considered are those that are proposed and explicitly stated.

Therefore, the possible alternatives for the region shown in Figure 12.1 may be a series of multiple purpose reservoirs, power plants, waste treatment plants, levees, and irrigation canals, as shown in Figure 12.2. A preliminary screening is now performed to determine if each component of the system is technologically and economically feasible. This may involve such things as checking to see if the geological conditions are such that a dam can be constructed at the chosen site, whether there would be enough stream flow to drive a power generator, whether environmental conditions will permit the construction of a power plant, and so on. This screening may thus reduce the number of possible alternatives that have to be evaluated.

The next step is to develop a model of the system that might be used in an analysis procedure that will aid in determining the components to be included in the final solution and their size. In order to develop such a model, data must be acquired for the stream flow, water quality, and projections for water use by individuals and industries, crop water requirements, and power use.

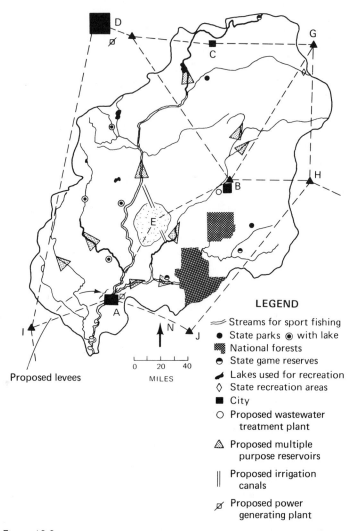

Figure 12.2

Alternatives Proposed for Solution to Problems from a Regional View

One possible approach might be to develop a linear programming model of the system. This problem might be stated as:

$$\text{Max } Z = B(E) + B(F) + B(IR) + B(W) - C(E) - C(F)$$
$$- C(IR) - C(W) - C(T) - C(R)$$

where

$B(E)$ = dollar benefits for power production

$B(F)$ = dollar benefits for flood control

$B(IR)$ = dollar benefits for irrigation

$B(W)$ = dollar benefits for water supply

$C(E)$ = dollar costs for power production exclusive of reservoir

$C(F)$ = dollar costs for flood control exclusive of reservoir

$C(IR)$ = dollar costs for irrigation exclusive of reservoir

$C(W)$ = dollar costs for water supply exclusive of reservoir

$C(T)$ = dollar costs for treatment plants exclusive of reservoir

$C(R)$ = dollar costs for reservoir

This criterion function must be maximized subject to the constraints on the system. The following are representative of the constraint relationships that must be developed.

1. Continuity at reservoirs for any time period:

   outflow $\leq$ inflow + amount in storage

   $$Q_0 \leq I + S$$

2. Storage capacity cannot be exceeded:

   $0 \leq$ amount in storage $\leq$ max capacity of reservoir

   $$0 \leq S \leq V$$

3. Flow through turbines:

   flow through turbines $\leq$ capacity of turbines

   $$Q_t \leq Q_{\max}$$

4. Water quality:

   water quality $\geq$ min acceptable water quality

   $$X_q \geq X_{\min}$$

In addition, there may be budgetary constraints; i.e., only so much money is available for developing the system.

After the model has been solved, a sensitivity analysis may be performed to determine the shadow prices that will provide an estimate of the marginal cost of satisfying the water quality standard. This will then allow the decision-maker to make a subjective judgment whether it might be appropriate to try a different standard.

It may also be desirable to state a minimum level for flood protection or water supply. Only what is explicitly stated in the model will be revealed by the answer obtained with the model. However, implicit assumptions may affect the answer; i.e., a constraint such as water quality may increase the cost several fold over that for a slightly lower quality.

A relationship must be established for the benefits and costs vs. size for each of the components in the system in order to perform an analysis such as this.

12.4 THE COMPONENT VIEW

Once the overall system and problem have been defined, each component can be represented as a subsystem (or a system in its own right). The goals of each subsystem are to contribute toward the objectives of the total system. That is, one objective of the total system may be to provide flood control that may then become the goal for the reservoir system component. Its objective is to reduce the outflow during a flood period, and the criteria that may be used to determine how efficiently it achieves this objective may be measured in terms of dollars.

The complete specification of a dam or reservoir requires determining a large number of attributes. However, during the initial formative design stages, the engineer can focus on a limited number of variables that relate to the reservoir height, storage, and its inflow and discharge characteristics.

Instead of finding the size of the dam specifically as an isolated parameter, however, the engineer is interested in determining a relationship between the size of the dam and its benefits and costs such that this relationship can be used in the criterion function. This might be done by using a dynamic programming approach as discussed in Chapter 11 for several different sizes of the reservoir. Alternatively, the engineer may develop linear programming or simulation models.

In order to provide the relevant technical data, the engineer needs hydrologic data for the river system for each seasonal period of the year. He also needs projections of social, technological (power, ecology, etc.), and use (drinking, recreation, etc.) requirements that may develop for the component. The engineer must develop relationships of total costs vs. reservoir size from estimations of cost based on similar projects using historical costs updated by price index figures and methods, as discussed in Chapter 9.

The initial planning stages for specific facilities necessarily focus on engineering considerations and requirements. However, once these broad details have been established, further planning must be compatible with the problems and resources available for actually implementing the facility.

Although topographic and geological conditions at the dam site may influence the selection of a specific dam type, the availability, quality, and economics of procuring suitable aggregates may decide whether the dam will be entirely concrete or dominantly rock or earth, with the minimal use of concrete in a spillway and apron.

If the component under consideration is the reservoir that includes as one of its purposes the storing of water to provide water supply to Town *B*, then one of its components will be a distribution system that will allow the water to be transported from the reservoir to the town and then to the users in the town. Although this is a part of the overall regional system, it may be the sole responsibility of Town *B* to see that it is implemented. Thus, it may be studied almost independently of the larger system by an engineer for the town.

12.5 IMPLEMENTATION PROBLEMS

Once a project has been defined and authorized for construction, the project engineer must set up a planning and scheduling CPM model for managing and controlling the project. For the reservoir component in the regional water system discussed above, he will develop models similar to those outlined in Chapters 6 and 10.

For each of the project activities, he must establish the necessary resources and initiate the material flows required to complete the activities. One particular project activity in the construction plan for the reservoir in Chapter 6 is "construct concrete spillway," which is a concreting operation of considerable size, since it costs $9 million and continues over a period of 20 months. The engineer is, therefore, faced with the problems of establishing the material flows of aggregate, sand, cement, and additives, and locating and constructing a complete concrete batching plant. He must determine its capacity, stockpile sizes, and so on, to insure uninterrupted production, as well as define the concrete delivery and placement processes.

The extent and use of concrete in the reservoir will have been decided by the locational characteristics and potential and economic sizes of suitable quarries. The financial costs associated with installing fixed or mobile rock crushing plants, plus the haulage fleets required to meet the concrete production requirements, must now be investigated. In some cases, cableways or belt conveyors may be more economical, especially in heavily broken country, and the engineer may carry out a decision tree analysis similar to those discussed in Chapter 7.

If suitable materials are not available in large quantities, the design will tend to minimize the concrete requirements. In some cases, it may become feasible to purchase ready-mixed concrete under contract from a neighboring town and design construction joints and schedule concrete pours to suit the locally available delivery fleet.

However, as the dam requires a large concrete spillway, the production and delivery of concrete becomes a major consideration in the design and economics of the facility. In these cases, considerable attention must be given to planning the entire concrete procurement process. Major considerations relate to the location of the concrete batching plant relative to the concrete site; its rated production capacity affects the duration of the concrete activity and depends on the total concrete quantities involved.

In order to establish the activity duration, the engineer may develop a simulation model for the batching plant operation and concrete truck fleet size similar to that developed in Chapter 8. Although the haulage distances for aggregate and sand are functions of the availability of suitable sites and economics, that for concrete involves additionally the important features of limited time before its first initial set, the quality deterioration associated with haulage, and its higher unit costs. Consequently, batching plants are usually located as close to the actual concreting site as possible or adjacent to the main materials delivery cable system. These requirements of continuous production at the concrete plant demand large inventory storage bins for sand, aggregate, and cement. In order to resolve locational and transportation problems, the engineer may develop linear programming models, as described in Chapter 5.

Defining the concrete procurement system requires considering many alternatives and system models. The use of cableways for materials delivery is very efficient, unaffected by terrain, weather conditions, and the type of material conveyed, but requires a heavy initial financial investment and steel tower construction. Road haulage systems have the advantage of cheaper initial construction costs and time and permit a staged development in haulage volume, but are either weather dependent or require heavy all-weather preparation. Although haulage roads may have limited final use, cableways have higher salvage potential.

Linear programming models can be developed to determine the optimal development of quarry sites and haulage distances and costs. Simulation studies can help determine stockpile sizes, inventory replacement policies, and the consequences associated with different batch plant sizes and concreting rates.

Extensive initial data must be collected before any decisive actions and designs can be undertaken. Data is required about topographic, soil, and geological conditions, road location, and dam site work areas and access. Extensive comparative estimates must be produced for a large number of alternative designs before final details are defined.

12.6 A MATHEMATICAL MODELING APPROACH TO THE CONCRETE BATCHING PLANT

The necessity for carefully considering the design and operation of a concrete batching plant is directly related to the magnitude of the concrete requirements of the project. If large quantities of concrete are required over a considerable period of time, opportunities may exist for both increasing production capacities and for realizing economies in concrete procurement. Engineers may then be called on to develop mathematical, economical, and simulation models of entire processes, and to develop management policies for their operation.

In general, mathematical models will be required for the following system features:

1. *System component and technology models.* These models will define the number, type, size, and characteristics of the plant entities involved in the concrete procurement process associated with the batching plant. In addition, the component technology models will define the transformations produced on the materials processed.

2. *The procurement process structure definition models.* These models define the routes taken by the various materials (sand, aggregates, cement, additives, concrete, etc.) through the various system components of the batching plant. Depending on the issues involved, the routes will be defined from sources (quarries, stockpiles, factories, bins, etc.) to the final destination (loading hopper, ready-mix truck, concrete spillway and apron, etc.).

3. *The management policy models.* These models will define the various actions to be taken when certain conditions exist, or provide criteria to select an alternative when several options become available. The models should focus on inventory features of material stockpiles, initiating and terminating concrete mixing, the rates at which quarries will be excavated and stockpiles replenished, and so on.

4. *System response models.* These models will determine the characteristics of the total procurement system in the batching plant that address the problems under consideration. Some will focus on the time characteristics of the system, such as the time response for changing concrete mixes, to deplete and replenish stockpiles, and idle time. Some, however, will focus on management problems associated with production costs and the feasibility of additions and alterations to eliminate system bottlenecks or to increase productivity.

The development of formal mathematical models will then permit analysis of the batching plant by providing the basis for functional steps in a computer program. The following material is intended to illustrate an approach to the mathematical modeling of the batching plant but does not attempt to either justify the level of modeling or present a complete model.

The concrete batching plant consists of a set of hardware component items connected by a system of belt conveyors, feed chutes, and hoppers. A typical layout for a plant is shown in Figure 12.3.

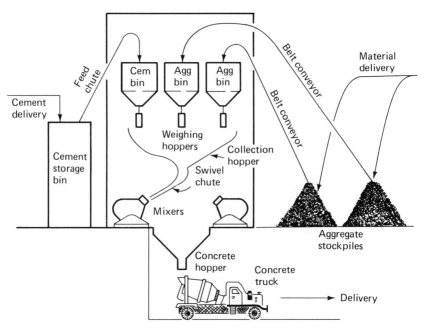

Figure 12.3

Concrete Plant Layout and Operations

The plant components can be described in symbolic terms with associated vectors of attributes to describe the physical and functional characteristics of the component. Thus, for example, the cement storage bin can be denoted by the symbol CEMENTSTORAGEBIN. Various attributes of the bin might be of interest; namely, its physical volume, the full capacity cement load, and the current status of the bin in terms of its cement content. These attributes can then either be assigned positional entries in a $N \times 1$ vector called CEMENTSTORAGEBIN(N) or given unique symbolic descriptors. The concrete plant mixer will have a capacity MIXVOLUME, a loading time MIXLOADTIME, a mixing time MIXTIME, and an emptying time, MIXDUMPTIME. The mixer times may be assumed to be constants or they may be probabilistic time distributions depending on the nature of the mixer operations and the level of modeling detail and problems requiring solutions. In a similar manner, each component shown in Figure 12.3 can be described so that a set of formal symbolic labels and vector attributes result.

The material flow processes are associated physically with the properties of the belt conveyors, feed chutes, and other mechanical parts. The flows can be expressed as time dependent rate equations with capacity limits. Thus, for example, the Ith aggregate belt conveyor BELTCONVEYOR(I) will have characteristics such as physical length, width, and inclination, as well as delivery rates in volume or weight units per unit of time as functions of belt speed, and the available transmission horsepower. Due to changing efficiency of equipment, aggregate bulking conditions on the belt, feed, and so on, the delivery rate may well be a variable with a specific probabilistic distribution that can be expressed in terms of the mean and standard deviation. Field observations on similar equipment may provide the raw data for defining specific model characteristics. For the cement and concrete delivery processors, discontinuous flow properties must be assumed as a function of time. Field data may be determined in a fashion similar to that outlined in Chapter 8.

Management policy statements can be related to the conditional occurrence of events or system states. Thus, for example, if the level of cement in the cement bin, CEMENTBINLEVEL(T), falls below a certain value, CEMENTBINLEVEL(A), the feed chute, CEMENTFEEDCHUTE, from the cement storage bin must be activated by the operator for a certain period of time, DELTATREFILL, to insure that the cement bin is almost filled. The policy statement for this situation may take the following form

IF (CEMENTBINLEVEL(T) $\leq$ CEMENTBINLEVEL(A)) THEN ACTIVATE
CEMENTFEEDCHUTE FOR TIME DELTATREFILL

Similar policy statements can be developed for each decision point in the batching plant operation.

The concrete plant is intended to produce a total volume of concrete, TOTALCONCRETE, over a construction period of time OPERATIONLIFE. It receives orders of size ORDERSIZE for a mix type, such as MIXTYPE(I). In reality, ORDERSIZE is a function of time and the various concrete mix types require different proportions of aggregate, sand, and cement.

Once sufficient variables and items have been identified, a variety of relationships can be developed linking the symbolic labels into a system of functional equations.

Thus, for example, from continuity considerations,

$$\int_{T=0}^{T=\text{OPERATIONLIFE}} \text{ORDERSIZE}(T)dT = \text{TOTAL CONCRETE}$$

$$= \text{ESTIMATED TOTAL CONCRETE}$$

$$+ \text{ADDITIONAL CONCRETE}$$

The amount of detail required and the system model scope that is developed will be a function of the problem definition and objectives. Various alternatives exist for the model both in the level of detail and the manner in which the parameters are characterized (e.g., by deterministic or probabilistic times). The result of the modeling efforts becomes a formal statement of the system and the system problem.

An initial modeling segment for the mixer production time, MIXERPRODUCTIONTIME, for the Nth order of ORDERSIZE(N) may be given by

$$\text{MIXERPRODUCTIONTIME} = \frac{\text{ORDERSIZE}(N)}{\text{MIXVOLUME}} \times (\text{MIXLOADTIME}$$

$$+ \text{MIXTIME} + \text{MIXDUMPTIME}) + \text{IDLETIME}$$

where IDLETIME is the amount of time during the process time for the order that the batching plant is not being serviced by delivery trucks at the loading dock. Consequently, IDLETIME is a function of the truck fleet size, TRUCKFLEETSIZE, and characteristics of the delivery cycle and concrete pouring operation in the field.

The actual model that the engineer develops must be compatible with an available system analysis procedure and supported by the specific data relevant to the particular problem under consideration.

12.7 JUSTIFYING THE SYSTEMS APPROACH TO PROBLEMS

The engineer has professional obligations to both his client and society. As a professional, he must exercise judgment and make decisions for which he will be held responsible. In many cases, his decisions are not quantifiable and are, therefore, subjective, although at all times he must strive to best meet societal needs and develop objective rationales for his decision-making. The engineer has this professional responsibility at whatever level he occupies in the hierarchy of decision-makers.

The engineer must always take the initiative and approach problems in their environmental context and search for a number of alternative solutions before selecting the most attractive. The systems approach assists the engineer in posing the problem in greater depth than he might otherwise have realized and in examining the full spectrum of alternate solutions that may be applicable. In this way, he can develop methods that permit trade-offs between various system solution attributes within the context and value system of complex societal forces and, thus, achieve better all-around solutions.

Finally, he must realize that he will be held responsible to society for any defects or consequential impacts that develop from his activities and decisions. In this respect, his only defense can be that he is continuously abreast of the engineering state of the art. The systems approach to problems helps the engineer to anticipate a variety of viewpoints and requirements and, thus, to forestall or minimize criticisms of his efforts from any segment of society.

12.8 PROBLEMS

P12.1. The planning and design activities are illustrated in Figure 1.2. For each of the following project types identify the components of each activity and the time scale, magnitude of effort, and resources required for each activity:

1. Selecting and purchasing a piece of equipment,
2. Extending a utility service to a small subdivision,
3. Expanding a warehouse,
4. Constructing major regional engineering works such as an irrigation system, interstate highway, or a power distribution system.

P12.2. The concepts of systems engineering, problem definition, and the systems approach to social problems in which engineers play a prominent role are discussed in the *Consulting Engineers Journal*, 1968 (see Bibliography). These areas are urban renewal, urban transportation, water pollution, air pollution, and rural redevelopment. Select one of these problem areas and:

1. Comment on the manner in which the author portrays the problem environment and formulates the solution approach.
2. In the proposed solution approach where can an engineer work alone and where and with whom must he interact with persons from other disciplines?

P12.3. The engineer is usually not told how to use the various approaches and techniques discussed in the text. Because problems are generally complex and the portion of the problem addressed by a given technique is limited, the engineer is faced with deciding where he can use the different techniques and how he can integrate them into a study of the overall problem. It may be possible, for example, for one technique to be relevant at several levels of the overall problem. The manner in which the engineer visualizes what the problem actually is, defines it, and gives it some structure is the hard part of most practical engineering problems. This is engineering and requires judgment, study, and often several attempts before the real problem is solved.

There are several possible ways of approaching the following problem; however, the questions are arranged so as to stimulate your thinking concerning some of the possible ways in which previously discussed techniques might be used and incorporated into an overall study of the problem.

A large construction project is located in a remote area and requires the development of a self-contained community that will support 5,000 residents to ensure availability of labor. One aspect of the development and operation of this community is handling and processing solid wastes (i.e., garbage).

> 1. Assuming that you are the engineer responsible for designing and planning this aspect of the community, pose the problem as a systems problem identifying components, identifying the requirements of that component, the basic issues associated with meeting those requirements, and their interaction within the system structure.
>
> 2. Linear graphs can be used in two different ways in the problem-solving process. There are certain problem-solving techniques associated with linear graphs such as cuts-sets or maximum flow-minimum cut theorem. In using these techniques, it is usually necessary to draw the graph even though it represents only the connectivity of the components. Sometimes this picture of the connectivity is all that can be obtained from the graph, since there is no solution technique associated with it for the particular problem. Nevertheless, the graph serves a very useful purpose in showing interdependence of the components. Identify and discuss where linear graph concepts could be used in this problem.
>
> 3. In order to use mathematical programming, the engineer must recognize problem areas over which he has some control and for which the possibility of obtaining an improved solution exists. The engineer must be able to formulate objectives and identify decision variables and constraints so that a mathematical programming formulation becomes possible. Where do you think

mathematical programming concepts could be usefully applied in this problem? What might be the decision variables and constraints involved? How would you go about quantifying the mathematical formulation?

4. An engineer must often make a decision now from among a set of alternatives that influence future actions and whose future outcomes are uncertain. Furthermore, he may have to document the basis for his decision. Among the decisions he may have to make are those regarding the type and depth of study required as well as which alternative is best. In the text, decision tree analysis, simulation, and state theory were proposed as possible techniques for use in decision-making. Discuss how these techniques might be used in this problem and how you would obtain the data necessary to apply these techniques. How would you handle situations in which no data exists and the need for immediate decisions precludes the possibility of gaining further data?

5. A basic problem in any project is identifying tasks, assigning resources to the tasks, and scheduling the work tasks. Project network models can be developed for areas with clearly defined sequential logic. Assigning resources and scheduling work tasks must often be made subjectively. You can solve the problem if you have certain information about the tasks, the order in which they can be performed, and the times required to do them. Some of this knowledge comes with experience on the job and cannot be taught. Most design and construction offices keep careful records of times required to do each job so that future estimates will be based on sound information. What is the relevance of project network models to this problem?

6. Indicate how the above studies can be integrated into an overall study of the solid waste· problem for the proposed community. Include:

    a. The type of information obtained from each step or study and how it would be used.

    b. A description of how these various phases or steps would interact; i.e., the way in which the results of one phase would affect the study being performed in another phase. Interaction may also result from the fact that another study is being performed even before results are available.

    c. The manner in which the validity of the results would be evaluated for both the individual steps as well as for the overall project. This part is designed to indicate what is involved in the complete planning and execution of a project. It is perhaps the most important part of the project because it integrates the isolated components of the approach into the problem.

P12.4. You are a plant engineer in charge of expanding the production facilities of the plant. How would you approach the expansion problem in order to increase manufactured output by

1. 10%,
2. 40%, and
3. 100%.

How would the project plans, focus, decision variables, and constraints change as the size of the expansion increases?

P12.5. Any work that an engineer performs is not very useful unless it is conveyed to other people. It may be necessary to transmit the information to sell the project or to get the project built after it is designed. The acceptability of the information is often influenced by the manner in which it is presented. The information is most often presented in the form of a report. The report outline might be the following:

1. Purpose of the report,
2. Need for the proposed alternative,
3. A general statement of the resources required by the alternative,
4. General design concepts,
5. Capital cost breakdown,
6. Operational cost breakdown,
7. Anticipated occupancy and income profile,
8. Work sheets.

An architectural-engineering firm is reporting to the client the results of their design study for a high-rise building. Comment on the format of the report. What do you think would be included in each of the sections of the report as outlined above? Where would you obtain the information you include? How might the report change to meet the requirements of decision makers at different levels?

P12.6. In order to achieve an optimal design, the decision-makers concerned with planning and design should explore the full range of choices open to them. There is a tendency to quickly accept a particular alternative solution and to concentrate only on the optimization of that one alternative. This usually results in a suboptimum solution. The technical problems of studying alternatives and the kinds of authority, flexibility, and motivation a water resource planning organization must have if it is to successfully analyze the appropriate alternatives in a problem are discussed by Davis (1968). In Chapters 1 and 2 Davis describes the problem environment of a water quality management problem and a proposed solution. Chapter 3 discusses the process of selecting alternatives for consideration. Chapters 9 and 10 discuss conditions that contribute to the incentive for planning organizations to increase their range of alternative solutions. Read these chapters and then:

1. List the major obstacles to the consideration of the full range of water quality management alternatives by the government agency in this case. Distinguish between those obstacles that were real constraints and those that were perceived as constraints.

2. How can some of these obstacles be removed?

3. How does the choice of goals, objectives, and criteria affect the range of alternatives to be analyzed?

P12.7. Large engineering projects in heavily populated and utilized areas inevitably affect many government, private, and local agencies and pose considerable site, access, and delivery problems. In order to cope with these issues in the decision process, extensive organizational structures and planning procedures are necessary.

1. Discuss the organizational structure and planning procedures that would have to be initiated in a project as large as the Verrazano-Narrows Bridge (*Civil Engineering*, 1964, see Bibliography).

2. Identify site, access, and delivery problems for this project.

P12.8. Community life raises many general problem areas that engineers must face; e.g., housing, food and water distribution, waste disposal, transportation, etc. The acceptance of particular solutions are influenced by the standards, magnitude, and level of social awareness that exists and that contributes to the environment in which the problem arises. The more complex the environment, the more forces, issues, and decision areas that must be considered. This requires a broader based interdisciplinary approach. Contrast your approach to the design and planning of a new integrated waste disposal system for a large existing city with the approach used in P12.3 for a remote area.

BIBLIOGRAPHY

THE SYSTEMS APPROACH TO ENGINEERING PROBLEMS

ANONYMOUS, "The Future of the Super Hi-Rise Building" *Modern Steel Construction*, First Quarter, 1972.

ASIMOW, M., *Introduction to Design*. Englewood Cliffs, N.J.: Prentice-Hall, Inc., 1962. A philosophy of engineering design is presented. This philosophy is comprised of three parts: a set of consistent principles; an operation discipline that leads to action; and a critical feedback apparatus that measures the advantages, detects the shortcomings, and illuminates the directions of improvement.

BOULDING, K. E., "General Systems As a Point of View," Chapter 2 in *Views on General Systems Theory*, ed. M. D. Mesarović. New York: John Wiley & Sons, Inc., 1964.

CHURCHMAN, C. W., *The Systems Approach*. New York: A Delta Book, Dell Publishing Company, Inc., 1968. Considers an approach that characterizes the nature of a system in such a way that decision-making can take place in a logical and coherent manner.

HEATHINGTON, K. W. AND R. B. BUNTON, "Systems Approach and the Civil Engineer," Engineering Issues, *Jour. Professional Activities, ASCE*, **97**, No. PP1, October 1971, 65–82, with discussion by T. C. KAVANAGH.

KATZ, D. L., R. O. GOETZ, E. R. LADY, AND D. C. RAY, *Engineering Concepts and Perspectives*. New York: John Wiley & Sons, Inc., 1968. A problem-oriented approach is used to stimulate engineering students to relate theory and formulations to a variety of real-world physical situations: automobile engines, rockets, electrical and control systems.

KAUFMANN, A., *The Science of Decision Making*. New York:

McGraw-Hill Book Company, 1968. Presents procedures by which the factors in a given situation can be analyzed and the result can be used to suggest the best course of action to be taken. The mathematical and graphical aspects of rather complex questions are alluded to, but only elementary problems are treated.

KRICK, E. V., *An Introduction to Engineering and Engineering Design* (2nd ed.). New York: John Wiley & Sons, Inc., 1969. This book presents a description of engineering practice, an introduction to important abilities of the engineer, and a motivating description of the fields in which he can profitably apply his talent.

MESAROVIĆ, M. D., D. MACKO, AND Y. TAKAHARA, *Theory of Hierarchical Multilevel Systems*. New York: Academic Press, Inc., 1970. In Part I of this book hierarchial systems are discussed in terms of levels of abstraction, levels of complexity of decision-making, and levels of priority in a multiunit decision system. A mathematical theory of coordination is developed in Part II.

"Systems Engineering Applied to Five Major Social Problems of Our Time," *Consulting Engineers Journal*, March 1968. Includes four papers on the principles of systems engineering, the systems team, systems techniques, and systems tools. Also includes papers by prominent politicians and consulting engineers on five major social problem areas: urban growth, urban transportation, water resources, air pollution, and rural development.

WILSON, W. E., *Concepts of Engineering Systems Design*. New York: McGraw-Hill Book Company, 1965. The text assumes that the engineering design of a system is of major interest to the student, and thereby introduces him to the concept of the profession and the function inherent in the design.

WOODSON, T. T., *Introduction to Engineering Design*. New York: McGraw-Hill Book Company, 1966. The book is written for the engineering student who is beginning his design experience, and it guides him through all stages of an authentic design project.

LINEAR GRAPH MODELING AND ANALYSIS OF SYSTEMS

AU, T. AND T. E. STELSON, *Introduction to Systems Engineering, Deterministic Models*, Chapter 6. Reading, Mass.: Addison-Wesley Publishing Company, 1969. Presents procedures for determining the maximum flows in networks with capacity constraints.

BUSACKER, R. G. AND T. L. SAATY, *Finite Graphs and Networks: An Introduction with Applications*. New York: McGraw-Hill Book Company, 1965. Graph theory is presented in the setting of a relatively informal discussion of central ideas that are amplified and illustrated by a variety of applications.

FRANK, H. AND L. T. FRISCH, "Network Analysis," *Scientific American*, July 1970. Describes the basic principles of graphical network analysis and their application in designing gas pipelines for off-shore drilling and in reliability analysis of electrical networks.

MARSHALL, C. W., *Applied Graph Theory*. New York: John Wiley & Sons, Inc., 1971. This is an introduction to graph theory and its application. Illustrative applications are given from the social sciences, physics, operations research, and related fields.

ORE, O., *Graphs and Their Uses*. New York: Random House, Inc., 1963. Presents an introduction to the kind of analyses that can be made by means of graphs and some of the problems that can be attacked by such methods. Only some of the simplest problems from graph theory are treated, so little technical knowledge is needed to understand the material.

SESHU, S. AND M. B. REED, *Linear Graphs and Electrical Networks*. Reading, Mass.: Addison-Wesley Publishing Company, 1961. A rigorous mathematical treatment of linear graph theory and its application to electrical networks. This book is aimed primarily at advanced graduate students who already have some fundamental knowledge of linear network analysis.

MATHEMATICAL MODELING OF ENGINEERING SYSTEMS

AU, T. AND T. E. STELSON, *Introduction to Systems Engineering, Deterministic Models*, Chapters 1 and 2. Reading, Mass.: Addison-Wesley Publishing Company, 1969. Presents the principles underlying different mathematical methods of analysis, with simplified examples for illustration.

BOUCHARD, H. AND F. H. MOFFITT, *Surveying* (5th ed.), Appendix. New York: International Textbook Company, 1966. The Appendix of this book presents a basic exposition of the principle of least squares and its applications in adjusting simple leveling triangulation and transverse nets.

BUSACKER, R. G. AND T. L. SAATY, *Finite Graphs and Networks*, Chapter 7. New York: McGraw-Hill Book Company, 1965.

Presents graphical and mathematical methods for analyzing network flows.

FENVES, S. J. AND F. H. BRANIN, "Network-Topological Formulation of Structural Analysis," *Jour. Structural Division, ASCE,* **89,** No. ST4, August 1963, 483–514.

FORD, L. R., JR., AND D. R. FULKERSON, *Flows in Networks.* Princeton, N.J.: Princeton University Press, 1962. Methods are presented for dealing with a variety of problems that have formulations in terms of flows in capacity-constrained networks. The problems discussed range from practical ones to more purely theoretical ones.

KESAVAN, H. K. AND M. CHANDRASHEKAR, "Graph-Theoretical Models for Pipe Network Analysis," *Jour. Hydraulics Division, ASCE,* **98,** No. HY2, February 1972, 345–364. Models based on concepts from linear graph theory are developed for analyzing nonlinear pipe networks.

MARSHALL, C. W., *Applied Graph Theory.* New York: John Wiley & Sons, Inc., 1971. This book is an introduction to graph theory and its application. Illustrative applications are given from the social sciences, physics, operations research, and related fields.

SESHU, S. AND M. B. REED, *Linear Graphs and Electrical Networks.* Reading, Mass.: Addison-Wesley Publishing Company, 1961. A rigorous mathematical treatment of linear graph theory and its application to electrical networks. This book is aimed primarily at advanced graduate students who already have some fundamental knowledge of linear network analysis.

WYMORE, A. W., *A Mathematical Theory of Systems Engineering: The Elements,* Chapter 3. New York: John Wiley & Sons, Inc., 1967. Presents the modeling of systems, with an interesting example on the mathematical modeling of an open pit copper mine.

OPTIMIZATION

ACKOFF, R. L., S. K. GUPTA, AND J. S. MINAS, *Scientific Method: Optimizing Applied Research Decisions,* Chapters 2 and 3. New York: John Wiley & Sons, Inc., 1962. Presents the meaning of an optimal solution and the approach to problem formulation.

HALL, W. A. AND J. A. DRACUP, *Water Resources Systems Engineering,* Chapters 3 and 4. New York: McGraw-Hill Book Company, 1970. This book presents fundamentals of the systems approach to complex water resources problems. Coverage includes the nature and objective functions of water resources systems; investment timing; large-scale, complex, multiple-purpose systems; ground-water systems; and water quality systems. Chapter 3 deals with the principles of systems analysis and Chapter 4 discusses the development of objective functions for water resources development.

HITCH, C. J., "On the Choice of Objectives in Systems Studies," in *Systems: Research and Design*, ed. Donald P. Eckman. New York: John Wiley & Sons, Inc., 1961.

KAUFMAN, A., *The Science of Decision Making*, Chapters 2 and 3. New York: McGraw-Hill Book Company, 1968. Chapter 2 suggests the use of the mathematical theory of sets as a basis for establishing the order of preference or value function. Chapter 3 discusses the use of linear programming models for optimization and the problems of suboptimization and sensitivity analysis.

WILDE, D. J. AND C. S. BEIGHTLER, *Foundations of Optimization*. Englewood Cliffs, N.J.: Prentice-Hall, Inc., 1967. This book presents a comprehensive, unified treatment of the field of optimization theory.

WOODSON, T. T., *Introduction to Engineering Design*, Chapters 13 and 15. New York: McGraw-Hill Book Company, 1966. Chapter 13 presents a discussion of criteria functions. Chapter 15 describes the basic principles and methods of optimization with an example on the optimum design of a bucket conveyor for a concrete mixing plant.

MATHEMATICAL PROGRAMMING

DANTZIG, G. B., *Linear Programming and Extensions*. Princeton, N.J.: Princeton University Press, 1963. A comprehensive treatment of the theoretical, computational, and applied areas of linear programming.

GASS, S. I., *Linear Programming* (3rd ed.). New York: McGraw-Hill Book Company, 1969. This is a basic presentation of the theoretical, computational, and applied areas of linear programming.

GREENBERG, H., *Integer Programming*. New York: Academic Press Inc., 1971. Presents theory and examples of integer programming and methods of solving practical problems.

IBM, *Mathematical Programming System/360 Version 2, Linear and Separable Programming—User's Manual* (3rd ed.), H20-0476-2. White Plains, N.Y.: IBM Corporation, 1969.

Lleywellyn, R. W., *Linear Programming*. New York: Holt, Rinehart and Winston, 1964. An introduction to the formulation and solution of linear programming problems. The model formulation and solution procedures for networks with multiple sources and sinks are presented clearly at an introductory level.

Wagner, H. M., *Principles of Operations Research: With Applications to Managerial Decisions*, Chapters 2 through 7, and 13. Englewood Cliffs, N.J.: Prentice-Hall Inc., 1969. An extensive coverage of the formulation and solution of linear models at an introductory level. Sensitivity testing and duality concepts are also covered.

Wilde, I. J. and C. S. Beightler, *Foundations of Optimization*. Englewood Cliffs, N.J.: Prentice-Hall Inc., 1967. This book presents a comprehensive, unified treatment of the field of optimization theory.

ORGANIZATIONAL NETWORKS

Antill, J. M. and R. W. Woodhead, *Critical Path Methods in Construction Practice* (2nd ed.) New York: John Wiley & Sons, Inc., 1970. Presents the concepts and procedures of this method of construction planning and project control. Discusses the use of CPM as a practical system for integrating project development and management.

Clough, R. H., *Construction Contracting* (2nd ed.), Chapter 12. New York: Wiley-Interscience, 1969. The chapter centers about a discussion of the critical path method and its associated applications of least-cost expediting and resource leveling.

Peurifoy, R. L., *Construction Planning, Equipment and Methods* (2nd ed.), Chapter 2. New York: McGraw-Hill Book Company, 1970. Discusses the application of the critical path method in job planning and management. A practical example on a highway project is used to illustrate the use of CPM to determine the duration of the project; to schedule materials, equipment and labor; and to schedule the amount and duration of the financing required for the project.

Shaffer, L. R., J. B. Ritter, and W. L. Meyer, *The Critical Path Method*. New York: McGraw-Hill Book Company, 1965.

Explains the purpose and function of the Critical Path Method. Emphasizes the application of the method through the use of practical examples taken from the construction industry.

Symposium, Verrazano-Narrows Bridge, *Civil Engineering*, **34,** No. 12, December 1964, also *Jour. Construction Division, ASCE*, **92,** No. CO2, March 1966.

WIEST, J. D. AND F. K. LEVY, *A Management Guide to PERT/CPM*. Englewood Cliffs, N.J.: Prentice-Hall Inc., 1969. Presents the basic ideas of PERT and CPM scheduling techniques and the variety of management problems to which they may be applied.

DECISION ANALYSIS

BARISH, N. N., *Economic Analysis for Engineering and Managerial Decision-Making*. New York: McGraw-Hill Book Company, 1962. Chapter 20 and 21. A discussion of different criteria and methods of analysis for decision-making under risk and uncertainty.

BENJAMIN, J. R. AND C. A. CORNELL, *Probability, Statistics, and Decisions for Civil Engineers*. New York: McGraw-Hill Book Company, 1970. A book on applied probability and statistics with the illustrations and problems taken from the civil engineering field.

FISHBURN, P. C., *Decision and Value Theory*, Chapters 2, 3, and 4. New York: John Wiley & Sons, Inc., 1964. These chapters deal with the decision structure and problem formulation, basic decision models, and the measurement of relative values (utility).

FISHBURN, P. C., *Utility Theory for Decision Making*. New York: John Wiley & Sons, Inc., 1970. Presents a unifying upper-level treatment of preference structures and numerical representations of preference structures.

RAIFFA, H., *Decision Analysis: Introductory Lectures on Choices under Uncertainty*. Reading, Mass.: Addison-Wesley Publishing Company, 1968. Decision tree is presented as an organizational scheme that can be used to help select the action to be taken when faced with a situation that requires a decision. The consequences of the chosen action may be uncertain. The Bayesian viewpoint, which uses both utilities and subjective probabilities, is utilized.

TRIBUS, M., *Rational Descriptions, Decisions and Designs*, Chapter 8. New York: Pergamon Press, 1969. In addition to decision tree and utility value, this chapter also deals with the decision problems relating to the design of experiments and with sequential testing.

SYSTEM SIMULATION

CHOW, V. T., "R & D of a Watershed Experimentation System," *JET Journal*, **15**, No. 6, February 1968, 11–13.

CHOW, V. T. AND B. C. YEN, "A Laboratory Study on the Effect of Moving Rainstorm on Surface Runoff," *Transactions*, American Geophysical Union, **49**, No. 1, March 1968, 168–169.

HUFSCHMIDT, M. M. AND M. B. FIERING, *Simulation Techniques for Design of Water-Resource Systems*. Cambridge, Mass.: Harvard University Press, 1966. Presents the various steps and procedures required to institute a simulation study of a water resource system, including procedures for collecting and organizing hydrologic and economic data, and for developing the necessary logic and detailed computer code using the Lehigh River Basin in Pennsylvania as an example.

IBM System/360 Operating System: FORTRAN IV(Z) Library Subprograms, Form C28-6596.

MAASS, A., M. M. HUFSCHMIDT, R. DORFMAN, H. A. THOMAS, JR., S. A. MARGLIN, AND G. M. FAIR, *Design of Water-Resource Systems*. Cambridge, Mass.: Harvard University Press, 1962. The results of a large-scale research program devoted to the methodology of planning and designing complex, multiunit, multipurpose water-resource systems. Discusses techniques of systems analysis appropriate for preliminary screening and the detailed analysis of alternatives.

MIDDLETON, J. T., "Planning Against Air Pollution," *American Scientist*, **59**, March–April 1971, 188–194.

BALINTFY, J. L., D. S. BURDICK, K. CHU, AND T. H. NAYLOR, *Computer Simulation Techniques*. New York: John Wiley & Sons, Inc., 1966. Discusses the rationale for computer simulation, the formulation of simulation models, as well as the design of simulation experiments. Also included are chapters on simulation languages and techniques for generating random numbers.

RAND Corporation, *A Million Random Digits with 100,000 Normal Deviates*. Glencoe, Ill.: The Free Press, 1955.

Random Number Generation and Testing, IBM Reference Manual C20-8011.

SOZEN, M. A., S. OTANI, P. GULKAN, AND N. N. NIELSEN, "The University of Illinois Earthquake Simulator," *Proceedings*, Fourth World Conference on Earthquake Engineering, Chile, January 1969.

TAKEDA, T., M. A. SOZEN, AND N. N. NIELSEN, "Reinforced Concrete Response to Simulated Earthquakes," *Jour. Structural Division, ASCE*, **96**, No. ST12, December 1970, 2557–2573.

SYSTEM PLANNING

BARISH, N. N., *Economic Analysis for Engineering and Managerial Decision-Making*. New York: McGraw-Hill Book Company, Inc., 1962. This is a technique-oriented presentation of the basic reasoning and methodology of economic analyses that are important in decision-making.

DE NEUFVILLE, R. and J. H. STAFFORD, *Systems Analysis for Engineers and Managers*. McGraw-Hill Book Company, Inc., 1972. Presents an extensive economic treatment of systems problems.

HILL, M., "A Goals-Achievement Matrix for Evaluating Alternative Plans," *Journal*, American Institute of Planners, **34**, No. 1, January 1968, 19–29. The goals-achievement matrix is postulated as a technique to be used in evaluating alternative plans. It is considered in terms of the requirements of the rational planning process.

JAMES, L. D. AND R. R. LEE, *Economics of Water Resources Planning*, Chapters 1 through 9, and 22. New York: McGraw-Hill Book Company, 1971. Presents the basic concepts of engineering economy, microeconomics, the criterion of economic efficiency, and financial analysis.

MASSE, P., *Optimal Investment Decisions: Rules for Action and Criteria for Choice*. Englewood Cliffs, N.J.: Prentice-Hall, Inc., 1962. A unifying treatment of the theory of investment, that includes linear programming, probability theory and stochastic processes, and applications of these techniques to problems in decision-making under both certainty and uncertainty.

THUESEN, H. G. AND W. J. FABRYCKY, *Engineering Economy* (3rd ed.). Englewood Cliffs, N.J.: Prentice-Hall, Inc., 1964. An introduction to the principles and techniques required for evaluating engineering alternatives in terms of worth and cost.

Water Resources Council, "Proposed Principles and Standards for Planning Water and Related Land Resources," *Federal Register*, Part II, **36,** No. 245, Washington, D.C., December 21, 1971, 24144–24194. Proposed objectives for water resources development and a method of analysis of alternative plans (the goals-achievement matrix).

PROJECT MANAGEMENT

FARKAS, L. L., *Management of Technical Field Operations*. New York: McGraw-Hill Book Company, 1970. Treats technical field operations as a special area of management with special practices, problems, and solutions.

FORRESTER, J. W., *Industrial Dynamics*. Cambridge, Mass.: The MIT Press, 1961. Presents a computer-based methodology for the modeling and analysis of industrial management systems.

HACKNEY, J. W., *Control and Management of Capital Projects*. New York: John Wiley & Sons, Inc., 1965. The author describes all aspects of controlling cost, time, and value and the interpersonal relationships that form the ever-present background for all capital projects.

HALPIN, D. W. AND R. W. WOODHEAD, "Heuristic Gaming Approach to Construction Management," *Preprint* 1550, ASCE Annual and National Environmental Engineering Meeting, St. Louis, Missouri, October 18–22, 1971.

HALPIN, D. W. AND R. W. WOODHEAD, "Flow Modeling Concepts in Construction Management," *Preprint 1618*, ASCE National Water Resources Engineering Meeting, Atlanta, Georgia, January 24–28, 1972.

RUBEY, H., J. A. LOGAN, AND W. W. MILNER, *The Engineer and Professional Management* (3rd. ed.). Ames, Iowa: The Iowa State University Press, 1970. Considers the fundamentals of management and its role in engineering and engineering services.

STARR, M. K., *Systems Management of Operations*. Englewood Cliffs, N.J.: Prentice-Hall, Inc., 1971. Considers the management of operations in an industrial environment. The interaction of flow, project, and scheduling models with inventory, quality, and facilities management reflect the operations manager's interrelations with marketing and finance.

STATE CONCEPTS OF SYSTEMS THEORY

ASHBY, W. R., *An Introduction to Cybernetics*, Chapter 3. New York: John Wiley & Sons, Inc., 1963. Presents the concept of system states clearly at an introductory level.

BELLMAN, R., *Dynamic Programming*. Princeton, N.J.: Princeton University Press, 1957. Presents an introduction to the mathematical theory of multistage decision processes (dynamic programming).

BELLMAN, R. E. AND S. E. DREYFUS, *Applied Dynamic Programming*. Princeton, N.J.: Princeton University Press, 1962. Describes how the theory of dynamic programming can be applied to the numerical solution of optimization problems in connection with satellites and space travel, the determination of trajectories, feedback control and servo-mechanism theory, inventory and scheduling processes, allocation of resources, and the determination of prices.

HEIDARI, M., V. T. CHOW, P. V. KOKOTOVIC, AND D. D. MEREDITH, "Discrete Differential Dynamic Programming Approach to Water Resources Systems Optimization," *Water Resources Research*, **7**, No. 2, April 1971, 273–282.

NEMHAUSER, G. L., *Introduction to Dynamic Programming*. New York: John Wiley & Sons, Inc., 1966. The theory and computational aspects of dynamic programming are presented at an introductory level.

WAGNER, H. M., *Principles of Operations Research with Applications to Managerial Decisions*, Chapters 8 through 12. Englewood Cliffs, N.J.: Prentice-Hall, Inc., 1969. An introduction to dynamic optimization model formulation and solution procedures with examples of dynamic programming for bounded and unbounded horizons.

SYSTEMS CONCEPTS IN ENGINEERING

Regional Planning Council, "A Consistent Trade-Off Approach to Rapid Transit System Planning," Baltimore, Maryland, February 1970. National Information Service Accession No. PB 192 692. Describes an application of linear programming to rapid transit system planning.

BRANDT, C. T., *A Systems Study of Soft Ground Tunneling*. Tulsa, Oklahoma: Fenix and Scisson, Inc., May 1970. National Technical Information Service Accession No. PB 194 769. Describes a study to investigate new ideas and radical concepts for soft ground tunneling. The project was supported jointly by the Office of High Speed Ground Transportation and by the Urban Mass Transportation Administration.

DAVIS, R. K., *The Range of Choice in Water Management.* Baltimore, Maryland: The Johns Hopkins Press, 1968. Economic-engineering analysis is used to explore alternatives and to provide the information needed to make comparisons and choices in the interests of serving society's preferences for the case of water quality planning for the Potomac estuary.

DAWES, J. H., "Tools for Water-Resource Study," *Jour. Irrigation and Drainage Division, ASCE,* **96,** No. IR4, December 1970, 403–424. Presents cost vs. size relationships suitable for preliminary planning studies for reservoirs, water-transmitting wells, pumps, municipal sewage treatment, and water pumping in Illinois.

DE NEUFVILLE, R., J. SCHAEKE, JR., AND J. H. STAFFORD, "Systems Analysis of Water Distribution Networks," *Jour. Sanitary Engineering Division, ASCE,* **97,** No. SA6, December 1971, 825–842. The systems analysis methodology is applied to the design and planning for the $1 billion Third City Tunnel for New York City.

GUN, W. A., "Airline Systems Simulation," *Operations Research,* **12,** No. 2, March 1964, 206–229.

KINDSVATER, C. E., ED., *Organization and Methodology for River Basin Planning.* Atlanta, Georgia: Georgia Institute of Technology, 1964. Proceedings of a seminar held at the Georgia Institute of Technology outlining the experiences of the U.S. Study Commission, Southeast River Basins, in organizing a staff and conducting comprehensive river basin planning.

MEREDITH, D. D. AND B. B. EWING, "Systems Approach to the Evaluation of Benefits from Improved Great Lakes Water Quality," *Proceedings,* 12th Conference of Great Lakes Research, International Association of Great Lakes Research, 1969, 843–870. Outlines a systems approach for evaluating benefits that would accrue due to an improvement in the quality of water in the Great Lakes. Presents a mathematical model that can be solved to determine the benefits from a change in water quality.

SHEMDIN, O. H., "River-Coast Interaction: Laboratory Simulation," *Jour. Waterways, Harbors and Coastal Engineering Division, ASCE,* **96,** No. WW4, November 1970, 755–766.

Appendix A
MATRIX ALGEBRA

Matrix algebra is a powerful mathematical language that has particularly significant applications in linear transformations and in manipulating linear equations. Whereas the full power of matrix algebra can be appreciated only by an in-depth study, the scope of this book limits this appendix to a mere exposition of the basic rules that are particularly important to beginning engineering students and are used liberally throughout this book to express mathematical equations in a condensed form.

A.1 DEFINITIONS

A matrix consists of rectangular arrays of elements, usually numerical, enclosed within a set of brackets. It has the following general format:

$$\begin{bmatrix} a_{11} & a_{12} & a_{13} & .. & a_{1n} \\ a_{21} & a_{22} & a_{23} & .. & a_{2n} \\ . & & & & . \\ . & & & & . \\ . & & & & . \\ a_{m1} & a_{m2} & a_{m3} & .. & a_{mn} \end{bmatrix}$$

where a_{ij} denotes one number (or one element) within the matrix. The first subscript, i, always denotes the row number in which the element a_{ij} lies; and the second subscript j always denotes the column of the element. Thus, the above matrix has m rows and n columns; and its dimension is said to be m by n. The dimension of the matrix can be written as $(m \times n)$ or simply (m, n).

Once the elements inside a matrix have been defined, the entire matrix may conveniently be assigned a name that consists of a single, bold face letter. For example, the above matrix may be assigned a name

A. Thus,

$$\mathbf{A} = \begin{bmatrix} a_{11} & a_{12} & a_{13} & .. & a_{1n} \\ a_{21} & a_{22} & a_{23} & .. & a_{2n} \\ \cdot & & & & \cdot \\ \cdot & & & & \cdot \\ a_{m1} & a_{m2} & a_{m3} & .. & a_{mn} \end{bmatrix}$$

The dimension of the matrix is sometimes written below its name to provide easy identification of its dimension; e.g., $\mathbf{A}_{(m,n)}$ or $\mathbf{A}_{(m \times n)}$.

Examples of matrices:

$$\mathbf{A}_{(3,3)} = \begin{bmatrix} 1 & 2 & 3 \\ 4 & 7 & 8 \\ 9 & 10 & 11 \end{bmatrix} \qquad \mathbf{B}_{(2,3)} = \begin{bmatrix} 1.1 & 2.3 & 4.7 \\ 8.9 & 7.6 & 3.3 \end{bmatrix}$$

$$\mathbf{X}_{(4 \times 1)} = \begin{bmatrix} x_1 \\ x_2 \\ x_3 \\ x_4 \end{bmatrix} \qquad \mathbf{D}_{(1 \times 5)} = \begin{bmatrix} 1 & 2 & 3 & 4 & 5 \end{bmatrix}$$

In the above examples, the elements of a matrix are real numbers or variables that may take on real numbers. There are several common forms of matrices. A matrix that has as many rows as columns is called a square matrix. For example, the above $\mathbf{A}_{(3 \times 3)}$ matrix has 3 rows and 3 columns, and is therefore a square matrix. A matrix that has only one column is called a column matrix, and a matrix that has only one row is called a row matrix. In the above examples, $\mathbf{X}_{(4 \times 1)}$ is a column matrix and $\mathbf{D}_{(1 \times 5)}$ is a row matrix.

A.2 OPERATING RULES

Equality

Two $(m \times n)$ matrices, say $\mathbf{A}$ and $\mathbf{B}$, are equal if, and only if, every element in $\mathbf{A}$ is equal to the corresponding element in $\mathbf{B}$; i.e.,

$$a_{ij} = b_{ij} \qquad \text{for all } i \text{ and } j$$

Example A.1

Let

$$\mathbf{A} = \begin{bmatrix} 1 & 2 & 3 \\ 4 & 7 & 8 \\ 9 & 10 & 11 \end{bmatrix} \qquad \mathbf{B} = \begin{bmatrix} 1 & 2 & 3 \\ 4 & 7 & 8 \\ 9 & 10 & 11 \end{bmatrix}$$

$$\mathbf{C} = \begin{bmatrix} 1 & 2 & 3 \\ 4 & 6 & 8 \\ 9 & 10 & 11 \end{bmatrix} \qquad \mathbf{D} = \begin{bmatrix} 1 & 2 & 3 & 0 \\ 4 & 7 & 8 & 0 \\ 9 & 10 & 11 & 0 \\ 0 & 0 & 0 & 1 \end{bmatrix}$$

Then, according to the definition of equality,

$$\mathbf{A} = \mathbf{B},$$
$$\mathbf{A} \neq \mathbf{C},$$
$$\mathbf{A} \neq \mathbf{D}.$$

The Addition Rule

The sum of two $(m \times n)$ matrices, $\mathbf{A}$ and $\mathbf{B}$, is defined as another $(m \times n)$ matrix, $\mathbf{C}$, such that elements in $\mathbf{C}$ are computed as

$$c_{ij} = a_{ij} + b_{ij} \qquad \begin{matrix} i = 1, 2, \ldots, m \\ j = 1, 2, \ldots, n \end{matrix}$$

That is,

$$\mathbf{C} = \begin{bmatrix} a_{11} + b_{11} & a_{12} + b_{12} & a_{13} + b_{13} & .. & a_{1n} + b_{1n} \\ a_{21} + b_{21} & a_{22} + b_{22} & a_{23} + b_{23} & .. & a_{2n} + b_{2n} \\ \cdot & & & & \cdot \\ \cdot & & & & \cdot \\ a_{m1} + b_{m1} & a_{m2} + b_{m2} & a_{m3} + b_{m3} & .. & a_{mn} + b_{mn} \end{bmatrix}$$

Similarly, the difference of two $(m \times n)$ matrices, $\mathbf{A}$ and $\mathbf{B}$, is defined as another $(m \times n)$ matrix, $\mathbf{C}$, the elements of which are:

$$c_{ij} = a_{ij} - b_{ij} \qquad \begin{matrix} i = 1, 2, \ldots, m \\ j = 1, 2, \ldots, n \end{matrix}$$

It follows from the definition that addition and subtraction can be performed only between two matrices having the same dimensions.

Example A.2

$$\mathbf{A} = \begin{bmatrix} 1 & 2 & 3 \\ 4 & 7 & 8 \\ 9 & 10 & 11 \end{bmatrix} \qquad \mathbf{B} = \begin{bmatrix} 4 & 7 & 9 \\ 3 & 2 & 1 \\ 6 & 5 & 3 \end{bmatrix}$$

Then,

$$\mathbf{A} + \mathbf{B} = \begin{bmatrix} 1+4 & 2+7 & 3+9 \\ 4+3 & 7+2 & 8+1 \\ 9+6 & 10+5 & 11+3 \end{bmatrix} = \begin{bmatrix} 5 & 9 & 12 \\ 7 & 9 & 9 \\ 15 & 15 & 14 \end{bmatrix}$$

and

$$\mathbf{A} - \mathbf{B} = \begin{bmatrix} 1-4 & 2-7 & 3-9 \\ 4-3 & 7-2 & 8-1 \\ 9-6 & 10-5 & 11-3 \end{bmatrix} = \begin{bmatrix} -3 & -5 & -6 \\ 1 & 5 & 7 \\ 3 & 5 & 8 \end{bmatrix}$$

Multiplying by a Scalar

When a matrix, $\mathbf{A}$, is multiplied by a scalar, c, the resultant matrix, $\mathbf{B}$, is obtained by multiplying every element in $\mathbf{A}$ by the scalar, c. That is,

$$\mathbf{B} = c\mathbf{A} = \begin{bmatrix} ca_{11} & ca_{12} & ca_{13} & . . & ca_{1n} \\ ca_{21} & ca_{22} & ca_{23} & . . & ca_{2n} \\ \cdot & & & & \cdot \\ \cdot & & & & \cdot \\ ca_{m1} & ca_{m2} & ca_{m3} & . . & ca_{mn} \end{bmatrix}$$

Consequences of Addition Rule and Multiplying by a Scalar

Let $\mathbf{A}$, $\mathbf{B}$, and $\mathbf{C}$ be three matrices having the same dimensions $(m \times n)$, and let a, b, and c be three scalar constants. Then, according to the above definitions, the following relationships can be easily proved:

1. $\mathbf{A} + \mathbf{B} = \mathbf{B} + \mathbf{A}$ (commutative under addition)
2. $\mathbf{A} + (\mathbf{B} + \mathbf{C}) = (\mathbf{A} + \mathbf{B}) + \mathbf{C}$ (associative under addition)
3. $c(\mathbf{A} + \mathbf{B}) = c\mathbf{A} + c\mathbf{B}$ (distributive under addition)
4. $(a + b)\mathbf{A} = a\mathbf{A} + b\mathbf{A}$
5. $a(b\mathbf{A}) = (ab)\mathbf{A}$

Proving these relationships is left to the student.

The Multiplication Rule

Let $\mathbf{C}$ be the matrix obtained by premultiplying the matrix $\mathbf{B}_{(r \times s)}$ by the matrix $\mathbf{A}_{(m \times n)}$; i.e.,

$$\mathbf{C} = \mathbf{A}_{(m \times n)} \cdot \mathbf{B}_{(r \times s)}$$

The product is commonly written as

$$\mathbf{C} = \mathbf{AB}$$

The multiplication rule states that this product is defined if, and only if, the first matrix, in this case $\mathbf{A}$, has as many columns as the number of rows in the second matrix, $\mathbf{B}$. That is, n must be equal to r. Furthermore, the elements, c_{ij}, of matrix $\mathbf{C}$ are computed from the following expression:

$$c_{ij} = \sum_{k=1}^{n} a_{ik}b_{kj}$$

The expression means that the c_{ij} element is obtained by multiplying the ith row of matrix $\mathbf{A}$ by the jth column of matrix $\mathbf{B}$.

Let

$$\mathbf{A} = \begin{bmatrix} a_{11} & a_{12} & a_{13} & .. & a_{1n} \\ a_{21} & a_{22} & a_{23} & .. & a_{2n} \\ \cdot & & & & \\ \cdot & & & & \\ a_{m1} & a_{m2} & a_{m3} & .. & a_{mn} \end{bmatrix} \qquad \mathbf{B} = \begin{bmatrix} b_{11} & b_{12} & b_{13} & .. & b_{1s} \\ b_{21} & b_{22} & b_{23} & .. & b_{2s} \\ \cdot & & & & \\ \cdot & & & & \\ b_{n1} & b_{n2} & b_{n3} & .. & b_{ns} \end{bmatrix}$$

Then,

$$\mathbf{C} = \mathbf{AB} = \begin{bmatrix} \Sigma a_{1k}b_{k1} & \Sigma a_{1k}b_{k2} & \Sigma a_{1k}b_{k3} & .. & \Sigma a_{1k}b_{ks} \\ \Sigma a_{2k}b_{k1} & \Sigma a_{2k}b_{k2} & \Sigma a_{2k}b_{k3} & .. & \Sigma a_{2k}b_{ks} \\ \cdot & & & & \cdot \\ \cdot & & & & \cdot \\ \Sigma a_{mk}b_{k1} & \Sigma a_{mk}b_{k2} & \Sigma a_{mk}b_{k3} & .. & \Sigma a_{mk}b_{ks} \end{bmatrix}$$

It follows from the above definition that the product, $\mathbf{C}$, has a dimension of $(m \times s)$. Thus, by writing the dimensions under the matrices in a matrix equation, it is possible always to keep a record of the dimensions of the new matrices; e.g.,

$$\mathbf{C}_{(m \times s)} = \mathbf{A}_{(m \times n)} \mathbf{B}_{(n \times s)}$$

Example A.3

Let

$$\mathbf{A}_{(2 \times 3)} = \begin{bmatrix} 1 & 3 & 7 \\ 2 & 4 & 3 \end{bmatrix} \qquad \mathbf{B}_{(3 \times 4)} = \begin{bmatrix} 2 & 3 & 7 & 1 \\ 1 & 2 & 3 & 2 \\ 4 & 1 & 3 & 6 \end{bmatrix}$$

Then,

$$\mathbf{C}_{(2 \times 4)} = \mathbf{A}_{(2 \times 3)} \cdot \mathbf{B}_{(3 \times 4)} = \begin{bmatrix} 1 & 3 & 7 \\ 2 & 4 & 3 \end{bmatrix} \begin{bmatrix} 2 & 3 & 7 & 1 \\ 1 & 2 & 3 & 2 \\ 4 & 1 & 3 & 6 \end{bmatrix}$$

$$= \begin{bmatrix} (2+3+28) & (3+6+7) & (7+9+21) & (1+6+42) \\ (4+4+12) & (6+8+3) & (14+12+9) & (2+8+18) \end{bmatrix}$$

i.e.,

$$\mathbf{C} = \begin{bmatrix} 33 & 16 & 37 & 49 \\ 20 & 17 & 35 & 28 \end{bmatrix}$$

Consequences of the Multiplication Rule

1. Matrix multiplication is associative and distributive if the dimensions of the matrices are compatible as required by the multiplication rule; i.e.,

$$(AB)C = A(BC)$$
$$(A + B)C = AC + BC$$

and
$$D(A + B) = DA + DB$$

2. Matrix multiplication is, in general, not commutative; i.e., $AB \neq BA$. For example, let

$$A = \begin{bmatrix} 2 & 3 \\ 4 & 1 \end{bmatrix} \qquad B = \begin{bmatrix} 7 & 1 \\ 2 & 3 \end{bmatrix}$$

Then,

$$A \cdot B = \begin{bmatrix} 2 & 3 \\ 4 & 1 \end{bmatrix} \begin{bmatrix} 7 & 1 \\ 2 & 3 \end{bmatrix} = \begin{bmatrix} 20 & 11 \\ 30 & 7 \end{bmatrix}$$

and

$$B \cdot A = \begin{bmatrix} 7 & 1 \\ 2 & 3 \end{bmatrix} \begin{bmatrix} 2 & 3 \\ 4 & 1 \end{bmatrix} = \begin{bmatrix} 18 & 22 \\ 16 & 9 \end{bmatrix}$$

3. $AB = 0$ does not imply that either $A = 0$ or $B = 0$. For example,

$$\begin{bmatrix} 1 & 1 \\ 2 & 2 \end{bmatrix} \begin{bmatrix} -1 & 1 \\ 1 & -1 \end{bmatrix} = \begin{bmatrix} 0 & 0 \\ 0 & 0 \end{bmatrix}$$

The Null and Identity Matrices

A matrix in which all the elements are zero is called a null matrix. It is equivalent to a zero in ordinary algebra. It can be easily proved that for any matrix A, then, $0A = A0 = 0$. However, as seen in the above paragraph, $AB = 0$ does not imply that either $A = 0$ or $B = 0$.

An identity matrix, universally denoted by the letter I, is a square matrix in which all the elements along the principal diagonal are equal to 1 and all the other elements are zero; i.e.,

$$I = \begin{bmatrix} 1 & 0 & 0 & 0 \\ 0 & 1 & 0 & 0 \\ 0 & 0 & 1 & 0 \\ 0 & 0 & 0 & 1 \end{bmatrix}$$

and $\mathbf{I}$ in this case is a (4×4) identity matrix. It is equivalent to unity (1) in ordinary algebra. It can be easily proved that for any matrix $\mathbf{A}$,

$$\mathbf{I}_{(m \times m)} \mathbf{A}_{(m \times n)} = \mathbf{A}_{(m \times n)} \mathbf{I}_{(n \times n)} = \mathbf{A}_{(m \times n)}$$

A.3   APPLICATIONS OF MATRIX ALGEBRA IN LINEAR TRANSFORMATION

In engineering applications, matrix algebra is useful as a shorthand mathematical language in manipulating linear equations. A complete set of thousands of simultaneous equations can be easily represented by a single matrix equation in one line. For example, consider the following set of linear equations that express the variables y_i as functions of the variables x_i's:

$$
\begin{aligned}
y_1 &= a_{11}x_1 + a_{12}x_2 + a_{13}x_3 + \ldots + a_{1n}x_n \\
y_2 &= a_{21}x_1 + a_{22}x_2 + a_{23}x_3 + \ldots + a_{2n}x_n \\
y_3 &= a_{31}x_1 + a_{32}x_3 + a_{33}x_3 + \ldots + a_{3n}x_n \\
&\quad \cdot \\
&\quad \cdot \\
y_m &= a_{m1}x_1 + a_{m2}x_2 + a_{m3}x_3 + \ldots + a_{mn}x_n
\end{aligned}
\tag{A.1}
$$

To express this system of equations in matrix notation, let

$$
\mathbf{Y}_{(m \times 1)} = \begin{bmatrix} y_1 \\ y_2 \\ \cdot \\ \cdot \\ y_m \end{bmatrix}, \qquad
\mathbf{A}_{(m \times n)} = \begin{bmatrix} a_{11} & a_{12} & a_{13} & .. & a_{1n} \\ a_{21} & a_{22} & a_{23} & .. & a_{2n} \\ \cdot & & & & \cdot \\ \cdot & & & & \cdot \\ a_{m1} & a_{m2} & a_{m3} & .. & a_{mn} \end{bmatrix}
$$

and

$$
\mathbf{X}_{(n \times 1)} = \begin{bmatrix} x_1 \\ x_2 \\ x_3 \\ \cdot \\ \cdot \\ x_n \end{bmatrix}
\tag{A.2}
$$

Then, the above equations can be simply stated as

$$\mathbf{Y}_{(m \times 1)} = \mathbf{A}_{(m \times n)} \mathbf{X}_{(n \times 1)} \tag{A.3}$$

In mathematical terms, Equations A.1 and A.3 are said to perform a linear transformation from a set of variables $(x_1, x_2, \ldots, x_n)$ to a second set of variables $(y_1, y_2, \ldots, y_n)$. The transformation process is linear because all the equations involve only first-order terms.

Consider a second transformation defined by the following set of equations:

$$z_1 = b_{11}y_1 + b_{12}y_2 + b_{13}y_3 + \ldots + b_{1m}y_m$$
$$z_2 = b_{21}y_1 + b_{22}y_2 + b_{23}y_3 + \ldots + b_{2m}y_m$$
$$\quad \cdot \qquad \cdot \qquad\qquad\qquad\qquad \cdot \qquad\qquad \text{(A.4)}$$
$$\quad \cdot \qquad \cdot \qquad\qquad\qquad\qquad \cdot$$
$$z_r = b_{r1}y_1 + b_{r2}y_2 + b_{r3}y_3 + \ldots + b_{rm}y_m$$

To express the z_i's as functions of the variables x_i's, the conventional approach in algebra is to substitute the Equations A.1 into Equations A.4 and collect the like terms. For example,

$$z_1 = b_{11}(a_{11}x_1 + a_{12}x_2 + a_{13}x_3 + \ldots + a_{1n}x_n)$$
$$+ b_{12}(a_{21}x_1 + a_{22}x_2 + a_{23}x_3 + \ldots + a_{2n}x_n)$$
$$+ \cdots\cdots\cdots\cdots\cdots\cdots\cdots\cdots\cdots$$
$$+ b_{1m}(a_{m1}x_1 + a_{m2}x_2 + a_{m3}x_3 + \ldots + a_{mn}x_n)$$

Collecting terms yields

$$z_1 = (b_{11}a_{11} + b_{12}a_{21} + \ldots + b_{1m}a_{m1})x_1$$
$$+ (b_{11}a_{12} + b_{12}a_{22} + \ldots + b_{1m}a_{m2})x_2$$
$$+ \cdots\cdots\cdots\cdots\cdots\cdots\cdots\cdots$$
$$+ (b_{11}a_{1n} + b_{12}a_{2n} + \ldots + b_{1m}a_{mn})x_n$$

To perform such an operation for every equation in (A.4) would be tedious and uninteresting. A much simpler procedure is offered by matrix algebra. Equation A.4 may be simply written as a matrix equation as follows:

$$\mathbf{Z}_{(r\times 1)} = \mathbf{B}_{(r\times m)} \mathbf{Y}_{(m\times 1)} \qquad \text{(A.5)}$$

Then, to perform the transformation process, Equation A.3 is simply substituted into Equation A.5 above; i.e.,

$$\mathbf{Z} = \mathbf{B}(\mathbf{A}\mathbf{X}) = (\mathbf{B}\mathbf{A})\mathbf{X}$$

A.4 The TRANSPOSE of a MATRIX

Let

$$\mathbf{A}_{(m \times n)} = \begin{bmatrix} a_{11} & a_{12} & a_{13} & .. & a_{1n} \\ a_{21} & a_{22} & a_{23} & .. & a_{2n} \\ a_{31} & a_{32} & a_{33} & .. & a_{3n} \\ . & & & & . \\ . & & & & . \\ a_{m1} & a_{m2} & a_{m3} & .. & a_{mn} \end{bmatrix}$$

The transpose of the matrix $\mathbf{A}$ is denoted as $\mathbf{A}^T$ and is defined as follows:

$$\mathbf{A}^T_{(n \times m)} = \begin{bmatrix} a_{11} & a_{21} & a_{31} & .. & a_{m1} \\ a_{12} & a_{22} & a_{32} & .. & a_{m2} \\ a_{13} & a_{23} & a_{33} & .. & a_{m3} \\ . & & & & . \\ . & & & & . \\ a_{1n} & a_{2n} & a_{3n} & .. & a_{mn} \end{bmatrix}$$

In general, the a_{ij}^t element in the transpose matrix is equal to the a_{ji} element in the original matrix; i.e.,

$$a_{ij}^t = a_{ji} \qquad \text{for all } i, j$$

The transpose matrix is obtained by flipping the original matrix so that each row becomes a column, and vice versa.

Example A.4

Let

$$\mathbf{A}_{(3 \times 4)} = \begin{bmatrix} 1 & 2 & 3 & 4 \\ 7 & 3 & 1 & 6 \\ 4 & 1 & 2 & 3 \end{bmatrix}$$

then

$$\mathbf{A}^T_{(4 \times 3)} = \begin{bmatrix} 1 & 7 & 4 \\ 2 & 3 & 1 \\ 3 & 1 & 2 \\ 4 & 6 & 3 \end{bmatrix}$$

The transpose matrix is often used in manipulating matrix equations. It can be easily proved that the following two relationships hold true for transpose matrices:

$$(\mathbf{A} + \mathbf{B})^T = \mathbf{A}^T + \mathbf{B}^T,$$

and

$$(\mathbf{A}\mathbf{B})^T = \mathbf{B}^T\mathbf{A}^T$$

Notice that the transpose of the product of two matrices is equal to the product of the transpose of the two matrices taken in *reverse* order.

A.5   The INVERSE of a MATRIX

The inverse of a square matrix, $\mathbf{A}$, is defined as that matrix $\mathbf{A}^{-1}$ such that

$$\mathbf{A}^{-1}\mathbf{A} = \mathbf{A}\mathbf{A}^{-1} = \mathbf{I} \tag{A.6}$$

Such an inverse exists if, and only if, the determinant of the matrix $\mathbf{A}$ is not equal to zero; i.e., $|\mathbf{A}| \neq \mathbf{0}$.

The inverse matrix is particularly useful in solving simultaneous equations. For example, let the following matrix equation represent n simultaneous equations, involving n variables:

$$\mathbf{A}_{(n \times n)}\mathbf{X}_{(n \times 1)} = \mathbf{C}_{(n \times 1)} \tag{A.7}$$

where

$$\mathbf{X}_{(n \times 1)} = \begin{bmatrix} x_1 \\ x_2 \\ x_3 \\ \cdot \\ \cdot \\ x_n \end{bmatrix} \quad \text{is a matrix of the } n \text{ variables}$$

Multiplying both sides of the equation by $\mathbf{A}^{-1}$,

$$\mathbf{A}^{-1}\mathbf{A}\mathbf{X} = \mathbf{A}^{-1}\mathbf{C}$$

i.e.,

$$\mathbf{I}\mathbf{X} = \mathbf{A}^{-1}\mathbf{C}$$

$$\therefore \qquad \mathbf{X} = \mathbf{A}^{-1}\mathbf{C} \tag{A.8}$$

Equation A.8 thus expresses a solution to Equation A.7. Although in practice a set of simultaneous equations is rarely solved by computing the inverse matrix as required in Equation A.8, this equation offers a simple means of deriving an expression for the unknowns. For example, let

$$\mathbf{Z}_{(r \times 1)} = \mathbf{B}_{(r \times n)}\mathbf{X}_{(n \times 1)} \tag{A.9}$$

Then, the solution for Z is simply obtained by substituting Equation A.8 into Equation A.9 as follows:

$$\mathbf{Z} = \mathbf{B}(\mathbf{A}^{-1}\mathbf{C}) = (\mathbf{B}\mathbf{A}^{-1})\mathbf{C}$$

Nevertheless, computing the inverse of a matrix is a common occurrence in solving engineering problems with the help of electronic computers. Consequently, many numerical techniques have been devised to perform the computation, and computer subroutines to perform such an operation are readily available from the computing centers.

A.6 COMPUTING THE INVERSE MATRIX

Consider the following (3×3) matrix:

$$\mathbf{A}_{(3 \times 3)} = \begin{bmatrix} 2 & 2 & 3 \\ 1 & 3 & 1 \\ 4 & 1 & 2 \end{bmatrix}$$

The following procedure is one variation of the Gauss Elimination Method.

Step 1: Combine the given matrix with a unit matrix $\mathbf{I}$ of the same dimension as follows:

$$\begin{bmatrix} 2 & 2 & 3 & \vdots & 1 & 0 & 0 \\ 1 & 3 & 1 & \vdots & 0 & 1 & 0 \\ 4 & 1 & 2 & \vdots & 0 & 0 & 1 \end{bmatrix}$$

Step 2: Divide every element in row 1 by a_{11}; i.e.,

$$\begin{bmatrix} 1 & 1 & \frac{3}{2} & \vdots & \frac{1}{2} & 0 & 0 \\ 1 & 3 & 1 & \vdots & 0 & 1 & 0 \\ 4 & 1 & 2 & \vdots & 0 & 0 & 1 \end{bmatrix}$$

Step 3: Multiply the elements in row 1 by $-a_{21}$ and add to the corresponding elements in row 2; i.e.,

$$\begin{bmatrix} 1 & 1 & \frac{3}{2} & \vdots & \frac{1}{2} & 0 & 0 \\ (1-1) & (3-1) & (1-\frac{3}{2}) & (0-\frac{1}{2}) & (1-0) & (0-0) \\ 4 & 1 & \frac{3}{2} & \vdots & 0 & 0 & 1 \end{bmatrix}$$

$$= \begin{bmatrix} 1 & 1 & \frac{3}{2} & \vdots & \frac{1}{2} & 0 & 0 \\ 0 & 2 & -\frac{1}{2} & \vdots & -\frac{1}{2} & 1 & 0 \\ 4 & 1 & 2 & \vdots & 0 & 0 & 1 \end{bmatrix}$$

Step 4: Multiply the elements in row 1 by $-a_{31}$ and add to the corresponding elements in row 3; i.e.,

$$\begin{bmatrix} 1 & 1 & \frac{3}{2} & \vdots & \frac{1}{2} & 0 & 0 \\ 0 & 2 & -\frac{1}{2} & \vdots & -\frac{1}{2} & 1 & 0 \\ (4-4) & (1-4) & (2-6)\cdot & (0-2) & (0-0) & (1-0) \end{bmatrix}$$

$$= \begin{bmatrix} 1 & 1 & \frac{3}{2} & \vdots & \frac{1}{2} & 0 & 0 \\ 0 & 2 & -\frac{1}{2} & \vdots & -\frac{1}{2} & 1 & 0 \\ 0 & -3 & -4 & \vdots & -2 & 0 & 1 \end{bmatrix}$$

The purpose of steps 2 to 4 is to reduce a_{11} to 1 and all other elements in column 1 to 0. The next sequence of steps is aimed at reducing element a_{22} to 1 and all other elements in column 2 to 0.

Step 5: Divide row 2 throughout by the latest value of a_{22}; i.e., 2.

$$\begin{bmatrix} 1 & 1 & \frac{3}{2} & \vdots & \frac{1}{2} & 0 & 0 \\ 0 & 1 & -\frac{1}{4} & \vdots & -\frac{1}{4} & \frac{1}{2} & 0 \\ 0 & -3 & -4 & \vdots & -2 & 0 & 1 \end{bmatrix}$$

Step 6: Multiply every element in row 2 by $-a_{12}$ and add to corresponding elements in row 1; and multiply every element in row 2 by $-a_{32}$ and add to corresponding elements in row 3;

$$\begin{bmatrix} (1-0) & (1-1) & (\frac{3}{2}+\frac{1}{4}) & \vdots & (\frac{1}{2}+\frac{1}{4}) & (0-\frac{1}{2}) & 0 \\ 0 & 1 & -\frac{1}{4} & \vdots & -\frac{1}{4} & \frac{1}{2} & 0 \\ 0 & (-3+3) & (-4-\frac{3}{4}) & \vdots & (-2-\frac{3}{4}) & (0+\frac{3}{2}) & \cdot(1+0) \end{bmatrix}$$

$$= \begin{bmatrix} 1 & 0 & \frac{7}{4} & \vdots & \frac{3}{4} & -\frac{1}{2} & 0 \\ 0 & 1 & -\frac{1}{4} & \vdots & -\frac{1}{4} & \frac{1}{2} & 0 \\ 0 & 0 & -\frac{19}{4} & \vdots & -\frac{11}{4} & \frac{3}{2} & 1 \end{bmatrix}$$

The next sequence of steps is aimed at reducing a_{33} to 1 and all other elements in column 3 to 0.

Step 7: Divide every element in row 3 by a_{33}; i.e., $-\frac{19}{4}$.

$$\begin{bmatrix} 1 & 0 & \frac{7}{4} & \vdots & \frac{3}{4} & -\frac{1}{2} & 0 \\ 0 & 1 & -\frac{1}{4} & \vdots & -\frac{1}{4} & \frac{1}{2} & 0 \\ 0 & 0 & 1 & \vdots & \frac{11}{19} & -\frac{6}{19} & -\frac{4}{19} \end{bmatrix}$$

Step 8: Multiply every element in row 3 by $-a_{13}$ and add to corresponding elements in row 1; and multiply every element in row 3 by $-a_{23}$ and add to corresponding elements in row 2.

$$\begin{bmatrix} (1+0) & (0+0) & (\frac{7}{4}-\frac{7}{4}) & \vdots & (\frac{3}{4}-\frac{7}{4}\times\frac{11}{19}) & (-\frac{1}{2}+\frac{7}{4}\times\frac{6}{19}) & (0+\frac{7}{4}\times\frac{4}{19}) \\ (0+0) & (1+0) & (-\frac{1}{4}+\frac{1}{4}) & \vdots & (-\frac{1}{4}+\frac{1}{4}\times\frac{11}{19}) & (\frac{1}{2}-\frac{1}{4}\times\frac{6}{19}) & (0-\frac{1}{4}\times\frac{4}{19}) \\ 0 & 0 & 1 & \vdots & \frac{11}{19} & -\frac{6}{19} & -\frac{4}{19} \end{bmatrix}$$

$$= \begin{bmatrix} 1 & 0 & 0 & \vdots & -\frac{5}{19} & \frac{1}{19} & \frac{7}{19} \\ 0 & 1 & 0 & \vdots & -\frac{2}{19} & \frac{8}{19} & -\frac{1}{19} \\ 0 & 0 & 1 & \vdots & \frac{11}{19} & -\frac{6}{19} & -\frac{4}{19} \end{bmatrix}$$

Finally, the inverse matrix is obtained from the former location of the I matrix; i.e.,

$$\mathbf{A}^{-1} = \begin{bmatrix} -\frac{5}{19} & \frac{1}{19} & \frac{7}{19} \\ -\frac{2}{19} & \frac{8}{19} & -\frac{1}{19} \\ \frac{11}{19} & -\frac{6}{19} & -\frac{4}{19} \end{bmatrix} = \frac{1}{19}\begin{bmatrix} -5 & 1 & 7 \\ -2 & 8 & -1 \\ 11 & -6 & -4 \end{bmatrix}$$

Since $\mathbf{A} \cdot \mathbf{A}^{-1} = 1$, the following computation can be performed as a check.

$$\mathbf{A}\mathbf{A}^{-1} = \begin{bmatrix} 2 & 2 & 3 \\ 1 & 3 & 1 \\ 4 & 1 & 2 \end{bmatrix} \cdot \tfrac{1}{19} \begin{bmatrix} -5 & 1 & 7 \\ -2 & 8 & -1 \\ 11 & -6 & -4 \end{bmatrix}$$

$$= \tfrac{1}{19} \begin{bmatrix} (-10 - 4 + 33) & (2 + 16 - 18) & (14 - 2 - 12) \\ (-5 - 6 + 11) & (1 + 24 - 6) & (7 - 3 - 4) \\ (-20 - 2 + 22) & (4 + 8 - 12) & (28 - 1 - 8) \end{bmatrix}$$

$$= \tfrac{1}{19} \begin{bmatrix} 19 & 0 & 0 \\ 0 & 19 & 0 \\ 0 & 0 & 19 \end{bmatrix} = \begin{bmatrix} 1 & 0 & 0 \\ 0 & 1 & 0 \\ 0 & 0 & 1 \end{bmatrix} \qquad \text{check.}$$

The above computational procedure can be readily generalized for square matrices for any size.

Appendix B
RANDOM NUMBERS

258164	244733	824904	959712	284925	062825
547250	466759	943814	751744	707634	376550
279794	797398	656465	505360	241001	256756
676883	778968	934335	028735	444391	538814
056700	668517	599657	172246	663342	229231
339846	006566	593875	032328	975552	373848
036783	039384	559225	193777	846672	240567
220480	236066	351556	161368	074279	441791
321406	414815	106967	967134	445197	647755
926274	486088	641104	796227	668169	882135
551342	913235	842276	771953	004479	286810
304312	473198	047928	626475	026876	718933
823825	835986	287273	754598	161107	308715
937351	010233	721707	522461	965570	850209
617730	061361	325338	131225	786849	095472
702187	367781	949838	786484	715749	572211
208356	204205	692568	713559	289632	429389
248744	223866	150708	276511	735843	573432
490798	341698	903251	657207	410058	436704
941463	047882	413364	938779	457579	617269
642372	286994	477391	626291	742379	699424
849870	720032	861112	753498	449229	191795
093443	315302	160820	515872	692334	149489
560052	889689	963853	091735	149304	895946
356517	332082	776563	549817	894838	369583
136699	990251	654104	295173	362940	215001
819290	934772	920183	769050	175190	288566
910170	602271	514838	609073	049977	729456
454833	609543	085541	650304	299551	371782
725920	653122	512693	897409	795288	228180
350587	914302	072686	378353	766325	367552
101159	479593	435653	267561	592743	202833
606294	874310	610972	603571	552441	215643
633650	239915	661686	617332	310901	292418
797598	437881	965626	699801	863313	752542
780166	624326	787185	194055	174009	510141
675692	741722	717763	163035	042897	057390
049565	445296	301705	977129	257123	343977
297081	668767	808201	856124	541013	061544
780488	008061	843715	130923	242413	368876
677624	048345	056556	784673	452850	210769
061141	289772	338980	702709	714037	263205
366460	736682	031592	211482	279375	577461
196284	415086	189369	267476	674370	460850
176396	487709	134951	603059	041642	761984
057200	922951	808817	614263	249601	565725
342842	531427	847407	681012	495933	396506
054743	184960	078683	083843	972241	376355
328117	108529	471591	502517	826831	255591
966490	650465	826352	011698	955365	531832
792363	898373	952494	070140	725692	187387
748793	384127	708486	420392	349224	123071
487669	302166	246104	519512	092989	737614
922713	810962	474975	113551	557332	420673
530000	860263	846638	680563	340216	521195
176412	155727	074070	078756	039000	123638
057296	933327	443946	472031	233763	741017
343418	593614	660673	828993	401014	441071
058191	557661	959553	968321	403375	643445
348781	342185	750789	803337	417523	856303
090331	050801	499633	814559	502313	131999
541400	304492	994413	881820	010477	791123

741361	587577	184041	904738	249258	043490
443133	521478	103020	422278	493873	260673
655800	125332	617446	530811	959896	562287
930349	751169	700488	181269	752846	369908
575766	501908	198174	086409	511964	216950
450692	008050	187719	517895	068314	300250
701101	048279	125060	103863	409449	799479
201847	289375	749535	622503	453928	791443
209734	734305	492120	730793	720497	743279
257001	400840	949389	379791	318087	454625
540279	402334	689874	276184	906380	724675
238011	411284	134560	655248	432116	343131
426470	464924	806470	927038	589773	056470
555938	786399	833339	555922	534638	338464
331860	713053	994366	331761	204207	028500
988926	273475	959432	988330	223879	170839
926828	639010	750068	923255	341774	023896
554667	829722	495313	533249	048337	143243
324241	972692	968524	195883	289723	858513
943259	829533	804554	173988	736391	145240
653138	971559	821856	042771	413344	870477
914393	822743	925550	256367	477271	216942
490140	930867	547007	536481	860395	300201
877588	578868	278336	215250	156522	799188
259562	469282	668147	290061	938092	789695
555628	812516	004348	738417	622172	732803
330002	869576	026091	425487	729812	391842
977790	211545	156401	550046	367921	348403
860090	267857	937368	296550	205045	088070
154690	605345	617831	777307	228898	527852
927116	627966	702794	658561	371856	163538
556390	763537	211993	946893	228627	980141
334568	576036	270543	674919	370231	874178
005151	452312	621442	044934	218884	239123
030902	710812	724437	269327	311841	433138
185230	260043	341699	614155	868948	595897
110145	558515	047892	680762	207778	571341
660151	347306	287055	079951	245282	424172
955428	081491	720398	479191	470046	542167
732061	488420	317493	871902	817095	249329
387397	927213	902818	225482	897017	494299
321763	556971	410766	351384	375999	962445
928409	338053	461820	105933	253457	768122
564139	026036	767792	634906	519039	603510
381010	156071	601535	805128	110720	616969
283489	935390	605132	825294	663595	697627
699030	605975	626689	946155	977067	181022
189434	631742	755882	670496	855755	084929
135342	786164	530157	018424	128713	509025
811161	711640	177351	110452	771429	050703
861454	265008	062927	661989	623332	303902
162865	588271	377162	967441	735763	821364
976108	525635	260428	798062	409580	922602
850006	150252	560818	782949	454713	529339
094259	900514	361107	692376	725202	172448
564942	396961	164207	149554	346286	033544
385822	379085	984151	896333	075379	201065
312328	271952	898211	371905	451791	205046
871865	629883	383157	228922	707687	228905
225262	775027	296355	371997	241317	371899
350066	644894	776133	229473	446283	228883
098036	864987	651527	375299	674678	371767

Appendix C
DISCOUNT FACTORS

5% Discount Factors

n	Single Payment		Uniform Series				n
	Compound Amount Factor CAF	Present Worth Factor PWSP	Compound Amount Factor USCA	Sinking Fund Factor SFP	Present Worth Factor PWUS	Capital Recovery Factor CRF	
	Given P to find F $(1+i)^n$	Given F to find P $\dfrac{1}{(1+i)^n}$	Given A to find F $\dfrac{(1+i)^n-1}{i}$	Given F to find A $\dfrac{i}{(1+i)^n-1}$	Given A to find P $\dfrac{(1+i)^n-1}{i(1+i)^n}$	Given P to find A $\dfrac{i(1+i)^n}{(1+i)^n-1}$	
1	1.050	0.9524	1.000	1.00001	0.952	1.05001	1
2	1.102	0.9070	2.050	0.48781	1.859	0.53781	2
3	1.158	0.8638	3.152	0.31722	2.723	0.36722	3
4	1.216	0.8227	4.310	0.23202	3.546	0.28202	4
5	1.276	0.7835	5.526	0.18098	4.329	0.23098	5
6	1.340	0.7462	6.802	0.14702	5.076	0.19702	6
7	1.407	0.7107	8.142	0.12282	5.786	0.17282	7
8	1.477	0.6768	9.549	0.10472	6.463	0.15472	8
9	1.551	0.6446	11.026	0.09069	7.108	0.14069	9
10	1.629	0.6139	12.578	0.07951	7.722	0.12951	10
11	1.710	0.5847	14.206	0.07039	8.306	0.12039	11
12	1.796	0.5568	15.917	0.06283	8.863	0.11283	12
13	1.886	0.5303	17.712	0.05646	9.393	0.10646	13
14	1.980	0.5051	19.598	0.05103	9.899	0.10103	14
15	2.079	0.4810	21.578	0.04634	10.380	0.09634	15
16	2.183	0.4581	23.657	0.04227	10.838	0.09227	16
17	2.292	0.4363	25.840	0.03870	11.274	0.08870	17
18	2.407	0.4155	28.132	0.03555	11.689	0.08555	18
19	2.527	0.3957	30.538	0.03275	12.085	0.08275	19
20	2.653	0.3769	33.065	0.03024	12.462	0.08024	20
21	2.786	0.3589	35.718	0.02800	12.821	0.07800	21
22	2.925	0.3419	38.504	0.02597	13.163	0.07597	22
23	3.071	0.3256	41.429	0.02414	13.488	0.07414	23
24	3.225	0.3101	44.500	0.02247	13.798	0.07247	24
25	3.386	0.2953	47.725	0.02095	14.094	0.07095	25
26	3.556	0.2812	51.112	0.01957	14.375	0.06956	26
27	3.733	0.2679	54.667	0.01829	14.643	0.06829	27
28	3.920	0.2551	58.400	0.01712	14.898	0.06712	28
29	4.116	0.2430	62.320	0.01605	15.141	0.06605	29
30	4.322	0.2314	66.436	0.01505	15.372	0.06505	30
35	5.516	0.1813	90.316	0.01107	16.374	0.06107	35
40	7.040	0.1421	120.794	0.00828	17.159	0.05828	40
50	11.467	0.0872	209.336	0.00478	18.256	0.05478	50
75	38.830	0.0258	756.594	0.00132	19.485	0.05132	75
100	131.488	0.0076	2609.761	0.00038	19.848	0.05038	100

6% Discount Factors

n	Single Payment		Uniform Series				n
	Compound Amount Factor CAF	Present Worth Factor PWSP	Compound Amount Factor USCA	Sinking Fund Factor SFP	Present Worth Factor PWUS	Capital Recovery Factor CRF	
	Given P to find F $(1+i)^n$	Given F to find P $\dfrac{1}{(1+i)^n}$	Given A to find F $\dfrac{(1+i)^n-1}{i}$	Given F to find A $\dfrac{i}{(1+i)^n-1}$	Given A to find P $\dfrac{(1+i)^n-1}{i(1+i)^n}$	Given P to find A $\dfrac{i(1+i)^n}{(1+i)^n-1}$	
1	1.060	0.9434	1.000	1.00001	0.943	1.06001	1
2	1.124	0.8900	2.060	0.48544	1.833	0.54544	2
3	1.191	0.8396	3.184	0.31411	2.673	0.37411	3
4	1.262	0.7921	4.375	0.22860	3.465	0.28860	4
5	1.338	0.7473	5.637	0.17740	4.212	0.23740	5
6	1.419	0.7050	6.975	0.14337	4.917	0.20337	6
7	1.504	0.6651	8.394	0.11914	5.582	0.17914	7
8	1.594	0.6274	9.897	0.10104	6.210	0.16104	8
9	1.689	0.5919	11.491	0.08702	6.802	0.14702	9
10	1.791	0.5584	13.181	0.07587	7.360	0.13587	10
11	1.898	0.5268	14.971	0.06679	7.887	0.12679	11
12	2.012	0.4970	16.870	0.05928	8.384	0.11928	12
13	2.133	0.4688	18.882	0.05296	8.853	0.11296	13
14	2.261	0.4423	21.015	0.04759	9.295	0.10759	14
15	2.397	0.4173	23.275	0.04296	9.712	0.10296	15
16	2.540	0.3937	25.672	0.03895	10.106	0.09895	16
17	2.693	0.3714	28.212	0.03545	10.477	0.09545	17
18	2.854	0.3503	30.905	0.03236	10.828	0.09236	18
19	3.026	0.3305	33.759	0.02962	11.158	0.08962	19
20	3.207	0.3118	36.785	0.02719	11.470	0.08719	20
21	3.399	0.2942	39.992	0.02501	11.764	0.08501	21
22	3.603	0.2775	43.391	0.02305	12.041	0.08305	22
23	3.820	0.2618	46.994	0.02128	12.303	0.08128	23
24	4.049	0.2470	50.814	0.01968	12.550	0.07968	24
25	4.292	0.2330	54.863	0.01823	12.783	0.07823	25
26	4.549	0.2198	59.154	0.01690	13.003	0.07690	26
27	4.822	0.2074	63.704	0.01570	13.210	0.07570	27
28	5.112	0.1956	68.526	0.01459	13.406	0.07459	28
29	5.418	0.1846	73.637	0.01358	13.591	0.07358	29
30	5.743	0.1741	79.055	0.01265	13.765	0.07265	30
35	7.686	0.1301	111.430	0.00897	14.498	0.06897	35
40	10.285	0.0972	154.755	0.00646	15.046	0.06646	40
50	18.419	0.0543	290.321	0.00344	15.762	0.06344	50
75	79.051	0.0127	1300.852	0.00077	16.456	0.06077	75
100	339.269	0.0029	5637.809	0.00018	16.618	0.06018	100

7% Discount Factors

n	Single Payment		Uniform Series				n
	Compound Amount Factor CAF	Present Worth Factor PWSP	Compound Amount Factor USCA	Sinking Fund Factor SFP	Present Worth Factor PWUS	Capital Recovery Factor CRF	
	Given P to find F $(1+i)^n$	Given F to find P $\dfrac{1}{(1+i)^n}$	Given A to find F $\dfrac{(1+i)^n-1}{i}$	Given F to find A $\dfrac{i}{(1+i)^n-1}$	Given A to find P $\dfrac{(1+i)^n-1}{i(1+i)^n}$	Given P to find A $\dfrac{i(1+i)^n}{(1+i)^n-1}$	
1	1.070	0.9346	1.000	1.00000	0.935	1.07000	1
2	1.145	0.8734	2.070	0.48310	1.808	0.55310	2
3	1.225	0.8163	3.215	0.31106	2.624	0.38105	3
4	1.311	0.7629	4.440	0.22523	3.387	0.29523	4
5	1.403	0.7130	5.751	0.17389	4.100	0.24389	5
6	1.501	0.6663	7.153	0.13980	4.766	0.20980	6
7	1.606	0.6228	8.654	0.11555	5.389	0.18555	7
8	1.718	0.5820	10.260	0.09747	5.971	0.16747	8
9	1.838	0.5439	11.978	0.08349	6.515	0.15349	9
10	1.967	0.5084	13.816	0.07238	7.024	0.14238	10
11	2.105	0.4751	15.783	0.06336	7.499	0.13336	11
12	2.252	0.4440	17.888	0.05590	7.943	0.12590	12
13	2.410	0.4150	20.140	0.04965	8.358	0.11965	13
14	2.579	0.3878	22.550	0.04435	8.745	0.11435	14
15	2.759	0.3625	25.129	0.03980	9.108	0.10980	15
16	2.952	0.3387	27.887	0.03586	9.447	0.10586	16
17	3.159	0.3166	30.840	0.03243	9.763	0.10243	17
18	3.380	0.2959	33.998	0.02941	10.059	0.09941	18
19	3.616	0.2765	37.378	0.02675	10.336	0.09675	19
20	3.870	0.2584	40.995	0.02439	10.594	0.09439	20
21	4.140	0.2415	44.864	0.02229	10.835	0.09229	21
22	4.430	0.2257	49.005	0.02041	11.061	0.09041	22
23	4.740	0.2110	53.435	0.01871	11.272	0.08871	23
24	5.072	0.1972	58.175	0.01719	11.469	0.08719	24
25	5.427	0.1843	63.247	0.01581	11.654	0.08581	25
26	5.807	0.1722	68.675	0.01456	11.826	0.08456	26
27	6.214	0.1609	74.482	0.01343	11.987	0.08343	27
28	6.649	0.1504	80.695	0.01239	12.137	0.08239	28
29	7.114	0.1406	87.344	0.01145	12.278	0.08145	29
30	7.612	0.1314	94.458	0.01059	12.409	0.08059	30
35	10.676	0.0937	138.233	0.00723	12.948	0.07723	35
40	14.974	0.0668	199.628	0.00501	13.332	0.07501	40
50	29.456	0.0339	406.511	0.00246	13.801	0.07246	50
75	159.866	0.0063	2269.516	0.00044	14.196	0.07044	75
100	867.644	0.0012	12380.633	0.00008	14.269	0.07008	100

8% Discount Factors

n	Single Payment		Uniform Series				n
	Compound Amount Factor CAF	Present Worth Factor PWSP	Compound Amount Factor USCA	Sinking Fund Factor SFP	Present Worth Factor PWUS	Capital Recovery Factor CRF	
	Given P to find F $(1+i)^n$	Given F to find P $\dfrac{1}{(1+i)^n}$	Given A to find F $\dfrac{(1+i)^n-1}{i}$	Given F to find A $\dfrac{i}{(1+i)^n-1}$	Given A to find P $\dfrac{(1+i)^n-1}{i(1+i)^n}$	Given P to find A $\dfrac{i(1+i)^n}{(1+i)^n-1}$	
1	1.080	0.9259	1.000	1.00000	0.926	1.08000	1
2	1.166	0.8573	2.080	0.48077	1.783	0.56077	2
3	1.260	0.7938	3.246	0.30804	2.577	0.38804	3
4	1.360	0.7350	4.506	0.22192	3.312	0.30192	4
5	1.469	0.6806	5.867	0.17046	3.993	0.25046	5
6	1.587	0.6302	7.336	0.13632	4.623	0.21632	6
7	1.714	0.5835	8.923	0.11207	5.206	0.19207	7
8	1.851	0.5403	10.637	0.09402	5.747	0.17402	8
9	1.999	0.5003	12.487	0.08008	6.247	0.16008	9
10	2.159	0.4632	14.486	0.06903	6.710	0.14903	10
11	2.332	0.4289	16.645	0.06008	7.139	0.14008	11
12	2.518	0.3971	18.977	0.05270	7.536	0.13270	12
13	2.720	0.3677	21.495	0.04652	7.904	0.12652	13
14	2.937	0.3405	24.215	0.04130	8.244	0.12130	14
15	3.172	0.3152	27.152	0.03683	8.559	0.11683	15
16	3.426	0.2919	30.324	0.03298	8.851	0.11298	16
17	3.700	0.2703	33.750	0.02963	9.122	0.10963	17
18	3.996	0.2503	37.450	0.02670	9.372	0.10670	18
19	4.316	0.2317	41.446	0.02413	9.604	0.10413	19
20	4.661	0.2146	45.761	0.02185	9.818	0.10185	20
21	5.034	0.1987	50.422	0.01983	10.017	0.09983	21
22	5.436	0.1839	55.456	0.01803	10.201	0.09803	22
23	5.871	0.1703	60.892	0.01642	10.371	0.09642	23
24	6.341	0.1577	66.764	0.01498	10.529	0.09498	24
25	6.848	0.1460	73.105	0.01368	10.675	0.09368	25
26	7.396	0.1352	79.953	0.01251	10.810	0.09251	26
27	7.988	0.1252	87.349	0.01145	10.935	0.09145	27
28	8.627	0.1159	95.337	0.01049	11.051	0.09049	28
29	9.317	0.1073	103.964	0.00962	11.158	0.08962	29
30	10.062	0.0994	113.281	0.00883	11.258	0.08883	30
35	14.785	0.0676	172.313	0.00580	11.655	0.08580	35
40	21.724	0.0460	259.050	0.00386	11.925	0.08386	40
50	46.900	0.0213	573.753	0.00174	12.233	0.08174	50
75	321.190	0.0031	4002.378	0.00025	12.461	0.08025	75
100	2199.630	0.0005	27482.879	0.00004	12.494	0.08004	100

INDEX

A

Accuracy
  in data collection, 93, 96–97, 203,
    216, 240, 332
  in modeling, 9, 15, 39, 80–82, 107,
    115, 157, 166, 173, 200–202,
    208–209, 250, 311, 332
Activity (*See also* CPM, Float, Organi-
    zation system)
  definition, 132, 165
  modeling, 37, 132, 165–166
    arrow form, 166, 168, 177–178
    bar form, 173, 176, 177, 179
    circle form, 166, 168
    project breakdown, 40, 166–167,
      177, 339–340, 347
    resources, 132, 166, 173, 177–182,
      285–286, 347
    scheduling, 3, 37, 40, 137–139, 141,
      165, 169–173, 287, 290–293,
      294, 309–324
Addition of matrices, 303, 364
Alternatives
  decision making aspects, 40, 65–67,
    197, 347
  evaluation (*See* Criteria, Decision
    analysis, Economic evaluation)
  generation, 13, 14, 21, 23, 32, 115,
    334–335
Analytical methods (*See* Mathematical
    modeling, Optimization)
Annual costs, 265, 267
Artificial variable (*See* Linear pro-
    gramming)

B

Bar chart (*See also* CPM)
  activity representation, 173, 176,
    292–293

modeling concepts, 176
uses
  cash requirements, 292–294
  project scheduling, 173, 176,
    177–182
Basic cut set (*See* Graph)
Basic feasible solution (*See* Feasibility,
    Linear programming)
Basic variable (*See* Linear program-
    ming)
Bayes theorem
  definition, 214–215
  use, in decision analysis, 215–219
  Venn diagram, 215
Behavior of systems (*See* Systems)
Benefit-cost (*See also* Economic
    evaluation)
  calculation, 266–267
  definition, 261–262
Bonds (*See also* Financial evaluation)
  issuance, 269–271
  need for, in project financing, 269
  repayment of loan, 272–273
  time profile of payments, 272
Branch (*See* Graph)
Branch-node incidence matrix (*See*
    Graph)
Branch-path matrix (*See* Graph)

C

Capital recovery factor, 265
Cash flow, 37, 262, 291–296 (*See also*
    Economic evaluation, Project
    management)
Chains (*See* Graph)
Chance node
  use, in decision analysis, 41, 200
  value, 206–208